FLEXBASING

Achieving Global Presence for Expeditionary Aerospace Forces

T0159555

Paul S. Killingsworth, Lionel Galway, Eiichi Kamiya, Brian Nichiporuk, Timothy L. Ramey, Robert S. Tripp, James C. Wendt

Prepared for the
United States Air Force

Project AIR FORCE
RAND

Approved for public release; distribution unlimited

The research reported here was sponsored by the United States Air Force under Contract F49642-96-C-0001. Further information may be obtained from the Strategic Planning Division, Directorate of Plans, Hq USAF.

Library of Congress Cataloging-in-Publication Data

Flexbasing : achieving global presence for expeditionary aerospace
forces / Paul S. Killingsworth ... [et al.].
   p. cm.
   " MR-1113-AF."
   Includes bibliographical references.
   ISBN 0-8330-2777-8
   1. United States. Air Force—Reorganization. I. Killingsworth,
Paul S.
UG633 .F554 2000
358.4' 00973' 0905—dc21   99-049554

Published 2000 by RAND
1700 Main Street, P.O. Box 2138, Santa Monica, CA 90407-2138
1333 H St., N.W., Washington, D.C. 20005-4707
RAND URL: http://www.rand.org/
To order RAND documents or to obtain additional information,
contact Distribution Services: Telephone: (310) 451-7002;
Fax: (310) 451-6915; Internet: order@rand.org

In August 1998, General Michael E. Ryan, the U.S. Air Force Chief of Staff, announced that the service would begin transforming itself into an Expeditionary Aerospace Force (EAF). The EAF reflects the Air Force vision of how it will organize, train, equip, and sustain itself in the 21st Century. This transformation is intended to meet the challenges of a new strategic environment, one that differs from the Cold War environment that shaped the Air Force as an institution for 50 years. The Air Force emerged from the Cold War with an organization and basing structure that was focused on the mission of containing the Soviet Union. Although there was a large permanent overseas presence, deployments of forces from the continental United States (CONUS) to overseas bases were rare, and when they occurred, as in exercises, large units of aircraft were moved from one well-stocked main operating base to another. The new security environment is placing quite different demands on the Air Force. There have been many deployments—some of them large, some of them small. The deploying forces are never able to rely on long-standing operations plans, but must rapidly tailor themselves to the demands of specific crises. These new security challenges are occurring in places that are distant from the Air Force structure of permanent bases, requiring long deployments of forces to relatively austere locations. Finally, the deployment burden has fallen unevenly, with some segments of the Air Force much more in demand for temporary overseas service than others. It has become clear to the Air Force leadership that the service needs to reshape itself to successfully operate in this new environment.

The Air Force has a history of reengineering itself to meet evolving strategic challenges, and the EAF concept is how the service will meet the latest challenges. The Air Force has made substantial progress in defining the command and control arrangements and organizational structures of the new EAF. However, these initiatives are not by themselves enough to make the Air Force expeditionary. This report focuses on another requirement—the need for a strategy to deploy and employ forces in the face of considerable uncertainty regarding overseas operating locations. The report proposes that this uncertainty can be managed by actively pursuing an overseas presence strategy based on maintaining high levels of logistical and operational flexibility, and outlines a number of far-reaching actions that the Air Force could take to leverage the advantages of aerospace power to gain overseas presence. These actions chiefly involve the establishment of a global logistical support system and provision of full-spectrum force protection for deployed forces. With the logistical wherewithal to deploy to a wide range of locations and a robust ability to protect itself, the Air Force will have the flexibility it needs to establish a global presence.

The research described in this report was conducted within the Aerospace Force Development Program of Project AIR FORCE, as part of a project entitled "Implementing an Effective Air Expeditionary Force." It was cosponsored by the Air Force Deputy Chief of Staff for Air and Space Operations (AF/XO) and the Deputy Chief of Staff for Installations and Logistics (AF/IL). The research should be of interest to Air Force leaders, operators, and planners who are implementing the EAF concept within the Air Force, as well as those in the field who are interested in understanding the strategic and historical background behind the movement to an EAF.

Project AIR FORCE

Project AIR FORCE, a division of RAND, is the Air Force federally funded research and development center (FFRDC) for studies and analyses. It provides the Air Force with independent analyses of policy alternatives affecting the development, employment, combat readiness, and support of current and future aerospace forces. Research is performed in four programs: Aerospace Force Development; Manpower, Personnel, and Training; Resource Management; and Strategy and Doctrine.

CONTENTS

FIGURES

TABLES

The U.S. Air Force has embarked on a process of reshaping itself to better meet the demands of the new strategic environment. This new environment presents challenges that are quite different from those the service faced when it came of age during the Cold War. In that struggle, the adversary was well known, and the theaters of operation were identified and defended with permanently stationed forces. Today's challenges are more diverse, and in many respects more unpredictable. There are both "pop-up" contingencies in places where the Air Force has rarely before operated and continuing "steady-state" regional security commitments far from any Air Force main operating base (MOB). This has forced a new mode of operation on the Air Force, one that has required frequent deployments of personnel and aircraft to austere forward operating locations. Not being structured to operate continuously in this way, the Air Force has had to pay a price for supporting these forward operations, a price that has been reflected in lower personnel retention rates and lower overall readiness. The service is responding to these challenges by reorganizing itself into an Expeditionary Aerospace Force (EAF). This reorganization represents an historic transition for the Air Force from a military service that has chiefly performed its mission by operating from MOBs to one that can quickly and easily project sizable forces overseas to austere and unanticipated locations, and sustain them there indefinitely.

To date, the Air Force has focused on the organizational aspects of this transition. It has designated ten Aerospace Expeditionary Forces (AEFs) that will rotate their availability for deployment and rapid response on a 15-month basis. This will allow the service to better manage the burden of temporary overseas deployments, while providing the warfighting commanders-in-chief (CINCs) with forces tailored to their needs. The Air Force, however, must go beyond revising its organizational structure if it is to become a truly expeditionary force. By definition, expeditionary forces need locations overseas from which to operate. Consequently, it is of great importance to the success of the EAF concept that the Air Force formulates and pursues a strategy aimed at providing the global presence it needs to perform its mission. The research described in this report examines the requirements for such a strategy.[1]

FLEXBASING: A STRATEGY FOR GLOBAL AEROSPACE PRESENCE

The Air Force emerged from the Cold War with an overseas basing structure that was centered on the two operational theaters of greatest concern at the time—Western Europe and Northeast Asia. In 1981, there were 41 of these bases, and today only 13 remain. Unfortunately, as shown by Figure S.1, these remaining bases are not well aligned to support operations in unstable regions around the world today.

The situation amounts to an expanded security perimeter for the United States. The perimeter is not only expanded geographically, as shown in the figure, but also expanded with respect to the nature and timing of the threats. In the past, great powers have met the challenges of extended frontiers by devising strategies that relied on operational flexibility, rapid mobility, agile logistics, global awareness, and advanced communications. These strategic capabilities, which are needed to manage the security environment of the next century, correspond closely with the capabilities and core competencies of aerospace power. To bring these competencies to bear, the Air Force needs to formulate and advocate a strategy that enables the projection of combat power to operating locations within the regions of

---

[1]In this report, the term Expeditionary Aerospace Force, or EAF, will refer to the current efforts to transform the Air Force as an institution into an expeditionary military force. The term Aerospace Expeditionary Force, or AEF, will refer to each of the ten groupings of forces the Air Force is using to schedule eligibility for overseas deployment. Forces that deploy overseas from these AEFs will be referred to as "AEF force packages," or sometimes as "expeditionary aerospace forces."

RAND*MR1113-S.1*

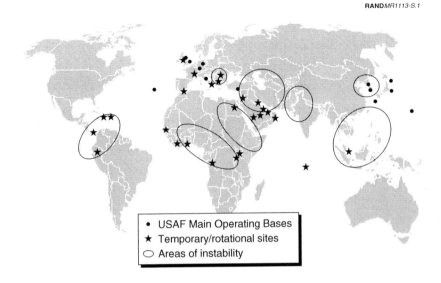

Figure S.1—USAF Overseas MOBs and Regional Instability, 1999

instability. Such a strategy is as important to the Air Force becoming expeditionary as the organizational and doctrinal efforts that are already under way.

We call our suggested strategy "flexbasing" because we believe that expeditionary forces can effectively manage uncertainty with respect to overseas operating locations by developing and maintaining a high degree of operational and logistical flexibility. This strategy replaces efforts to achieve an elusive assured access to specific overseas operating sites with the development and maintenance of a robust *capability* to deploy to and operate from a range of locations with widely varying characteristics. These locations could be allied military bases, international airports, or abandoned airfields. They could be relatively distant from the combat area, requiring the use of long-range strike capabilities, or they could be quite close, posing force protection challenges. The strategy can be implemented by applying the operational, logistical, and space competencies of the Air Force in the following ways:

- **Establish a global system of core support locations (CSLs), forward support locations (FSLs), and forward operating locations (FOLs).** The CSLs (normally the home bases of EAF units) and regionally located FSLs will be situated to support the rapid deployment of expeditionary aerospace power into a large number of possible FOLs in a region. The FSLs will usually be simply storage sites, but they could also provide regional maintenance facilities for ongoing operations in a theater. FSLs will also provide en route infrastructure for air mobility forces and beddown sites for bombers and enabling assets.

- **Develop and maintain a robust mix of both long- and short-range combat capabilities.** Expeditionary operations should not emphasize one capability over the other, but maintain a flexible balance. They must have the capability to project combat power whether or not close-in bases are immediately available. In addition, long-range operations will often help to enable later access to locations closer to the action during a crisis.

- **Develop space as an FSL supporting expeditionary operations.** Low-earth orbit is a regional support location for expeditionary deployments that is situated only 200 miles from any theater of operations. In addition, Air Force space operations represent a relative strength for the United States, and provide almost assured access. The Air Force should seek to place as many enabling assets as it can in earth orbit. Among the missions that are being considered for accomplishment from space are the Airborne Warning and Control System (AWACS) and Joint Surveillance and Tracking System (JSTARS) functions. The feasibility of accomplishing the Suppression of enemy air defenses (SEAD) and antitheater ballistic missile missions from space should also be investigated.

- **Advocate a global presence strategy such as flexbasing as a joint initiative.** A global support infrastructure to enable expeditionary operations would not support only Air Force operations. All services will be able to use FSLs to regionally locate support equipment and supplies. The maintenance of the flexbasing system will need to be established as a specific goal of the *shaping* aspect of U.S. military strategy.

- **Design and establish a global logistics/mobility support system for the EAF concept.** This system, set up on a worldwide basis, will provide the combat support flexibility that expeditionary aerospace forces will need to deploy anywhere within a region, and then to rapidly commence and sustain operations. Our research addressed this key near-term enabler of a global presence strategy in detail.

- **Provide full-spectrum force protection to deployed expeditionary forces.** Force protection is a fundamental requirement for expeditionary aerospace forces. Just as expeditionary forces must have long-range capabilities to operate effectively from distant locations, they must also have the capability to operate from locations that are possibly very forward. Robust force protection will lower the likelihood that enemy threats could prevent expeditionary forces from deploying to an FOL. This is another key enabler of flexbasing.

The combination of these initiatives will enable expeditionary aerospace forces to deploy to widely varying locations. They will not be dependent on access to any particular base, and will have potential access to many locations throughout a theater of operations. Figure S.2 shows a notional example of the flexbasing concept—as a globally planned system to support the projection of aerospace power. Additional analysis to develop this concept is under way in a series of follow-on studies of logistics, mobility, and base access.

A GLOBAL LOGISTICS/MOBILITY SUPPORT SYSTEM

We addressed the last two aspects of the flexbasing strategy in greater detail. The first of these was to provide expeditionary forces with a high level of mobility and logistics flexibility through the design of a globally planned combat support system that closely integrates logistics and mobility capabilities. In observing the early deployments of AEF force packages, we concluded that with today's support equipment and processes, the Air Force cannot achieve the very high levels of deployability it seeks for rapid-response deployments without prepositioning substantial amounts of infrastructure at the

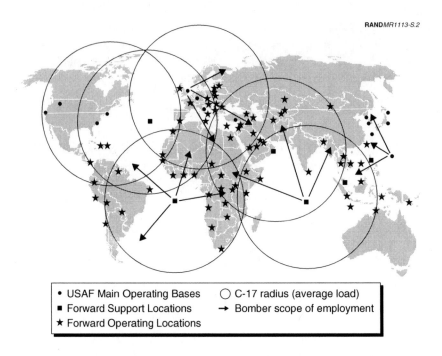

Figure S.2—Notional Flexbasing Concept

expected FOLs.[2] We also found that to have high confidence of gaining access to a prepositioned set, a number of these sets would need to be located throughout a given region, incurring additional support costs.

Our analysis indicated that positioning this infrastructure at FSLs and moving it to the FOLs with theater airlift represented a good compromise between cost and responsiveness. Generalizing this finding to support the goal of a worldwide EAF force projection capability led to the conceptualization of a global logistics/mobility support system with the following key elements:

---

[2]The benchmark for rapid AEF force package deployability is usually expressed as a maximum of 48 hours from the time the deployment order is given to the generation of the first combat sortie at the FOL.

- **Forward operating locations** represent a potentially large number of deployment sites throughout a theater. They will have varying levels of prepositioned U.S. infrastructure depending on the level of U.S. interest and the quality of the relationship with the host country. We found that FOLs with the highest responsiveness will require the most in-place resources, and consequently will be substantially more expensive than FOLs with less responsiveness.

- **Forward support locations** are regional support facilities outside of CONUS, located at sites with high assurance of access. They will be joint facilities, normally staffed at low levels. They will also take maximum advantage of host nation funding and commercially available products and services. The resources stored at FSLs will vary with the defense requirements of the region, but could include munitions, spare parts, war reserve materiel (WRM), and humanitarian supplies. We concluded that FSLs will be essential to affordably supporting rapid deployments within a region, as well as for sustaining the deployed forces. We also found that in many cases FSLs will be the best option for conducting intermediate-level maintenance on engines and avionics components. By centralizing these functions at FSLs and not deploying them forward with each AEF force package, we determined that substantial reductions in deployment footprint and maintenance manpower requirements could be achieved.

- **Core support location** facilities are usually CONUS and overseas MOBs that are the home bases for expeditionary forces. However, they could also be contractor facilities or military depots that provide various types of support to deployed forces, such as consolidated maintenance functions. In some cases, CSLs will serve as backups to the FSLs.

- **An air mobility network.** The transportation network will support the peacetime and crisis movement of equipment and personnel between the FOLs, FSLs, and CSLs. The air mobility system will enable the periodic deployments and redeployments of forward-based AEF force packages. In addition, it will support the upkeep and surveillance of the infrastructure placed throughout the flexbasing system and exercise the routes used for assured resupply during wartime.

- **A logistics command and control (C2) system** to facilitate decisionmaking and the flow of requirements information. Logistics C2 will also enable the system to react swiftly to rapidly changing circumstances.

The entire structure will need to be supported by a dynamic strategic planning process. This process must be informed by an analysis capability that can address issues such as what to preposition and where, the functions that should be performed at FSLs, and how many of each type of FOL should be set up. Decisions such as these must be made centrally for the entire system, so that mutual support between theaters can be leveraged and global transportation networks established. Centralized planning will be essential if the support system is to be affordable and sustainable over time.

FORCE PROTECTION FOR EXPEDITIONARY AEROSPACE FORCES

Although a highly capable logistics and mobility system will provide great flexibility of basing options, enemy threats could intervene to prevent AEF force packages from exploiting those options. FOLs with substantial levels of prepositioned materiel could have those resources denied to deploying expeditionary forces by credible enemy threats. Force protection is clearly partnered with logistics and mobility capabilities in enabling the flexbasing strategy. Together, the Air Force can deploy forces to wherever they are needed.

We examined a range of ground, air, chemical/biological, and information warfare threats to AEF force packages deployed at FOLs, at varying levels of intensity. Our purpose was to identify where the EAF concept needs to place its emphasis to achieve a deployment capability that is less constrained by concerns over the security of its forward bases. We identified four broad areas that need attention if expeditionary aerospace forces are to have the force protection capability they need.

Better Sensors and Firepower

Attacks on forward deployment locations could afford adversaries their best option for countering U.S. aerospace power. Security forces at forward bases need new sensors and weapons to protect deployed forces from a range of ground, air, and chemical attacks. For example, the standoff threat to forward bases from mortars or rockets launched from outside of the base perimeter is a serious one. To counter this threat, a tactical unmanned air vehicle (UAV) with an infrared sensor is needed. Additionally, counterbattery and counter-sniper technology should be fielded. Better sensors are also needed to detect and evaluate air traffic in the vicinity of the FOL, and for de-tecting chemical and biological contaminants in food and water.

Antitheater Ballistic Missile and Cruise Missile Capability

Today, the missiles fielded by most potential adversaries have poor accuracy and consequently little military value. Within the next decade, however, expeditionary forces will face theater ballistic mis-siles (TBMs) with substantially improved accuracy. In addition, the use of Global Positioning System (GPS) guidance systems and low observable technology will in all likelihood increase the threat of cruise missiles. Unless countermeasures are fielded, these weapons will circumscribe the range of possible FOLs available. As a result, the Air Force may be forced to rely solely on long-range weapon sys-tems, which would reduce its operational flexibility. The Patriot PAC-3 system will have an improved capability against both TBMs and cruise missiles, as will a number of follow-on systems. However, none of these systems is highly deployable. There is no easy counter to these threats that meets the needs of expeditionary aerospace forces. Until an effective and deployable system is available, the Air Force must manage the risk with a combination of deterrence, prepositioning of defensive systems, and lengthening of employment timelines to include the deployment of defenses.

Collective Protection Against Chemical/Biological (CB) Weapons

Like TBMs, CB weapons have the potential to greatly reduce the de-ployment flexibility of expeditionary aerospace forces. Although great progress has been made in enabling deployed forces to operate in limited CB environments, ultimately expeditionary forces will face adversaries with sophisticated military CB capabilities. To prepare for this threat, the Air Force needs to procure a deployable collective protection (COLPRO) capability. COLPROs could allow base operability while decontamination takes place. Without COLPROs, the only option for deployed expeditionary forces after a large CB attack would be to evacuate the base, leaving behind contaminated support equipment and shelters.

Evaluate the Threats to Reachback Capabilities

Expeditionary aerospace forces can significantly reduce their support footprint and enhance their effectiveness through communications "reachback" to rear areas for many command and control, intelligence, and planning functions. Smaller rapidly deploying and employing forces will especially need to leverage reachback for force protection and force enhancement. Denial of these capabilities through the use of communications jamming or information warfare (IW) attacks has the potential to substantially reduce the effectiveness of deployed forces. Although the degree to which this is a threat today is unknown, it is certain that it will increase in the future. The Air Force needs to evaluate the threat to its information flows and design appropriate measures to protect them.

FLEXBASING AND OPERATIONAL EFFECTIVENESS

We examined the effects of logistics support, force protection, and the flexbasing strategy on the generation of operational capability at forward locations by considering three cases involving the deployment of expeditionary packages of airpower.[3] The first, Case A, posited deployment to a well-stocked forward base under the threat of TBMs and CB agents. Case B looked at a deployment to a bare base with a similar high-intensity threat. Case C was a deployment to a moderately equipped base with a low threat. For each case, we considered the effect on operations of a regional forward support location at which to base sustaining supplies, bombers, and enabling assets such as tankers and AWACS. The measure of merit was the number of Joint Direct Attack Munitions (JDAMs) that could be

---

[3]Deployment of 12 F-15Es, 12 F-16CGs, 12 F-16CJs, and 6 B-1s was assumed. B-2s were employed directly from CONUS.

delivered over a two-week period. The results are shown below in Figure S.3.

As would be expected, an additional base in the theater resulted in more JDAMs being delivered in each case. Moreover, the FSL contributes substantially to the early application of force. In Case A, with a well-equipped base, only 168 JDAMs could be delivered in the first five days without access to an FSL, as opposed to 585 with an FSL. Case B indicates that FSLs contribute the most to operations in theaters without built-up FOLs. In this deployment to a bare base

RAND*MR1113-S.3*

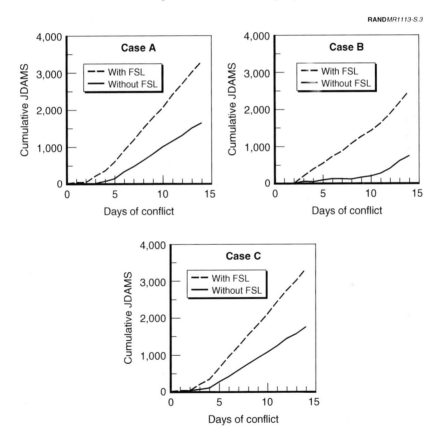

Figure S.3—Precision-Guided Munitions Delivered

without an FSL, 96 JDAMs were delivered during the first five days. With an FSL, 561 could be delivered. In this case, the results represent the difference between pin-prick strikes and an air campaign. In every case, the intensity of the strikes, as represented by the slopes of the lines, was significantly greater with access to an FSL. We found that the FSLs envisioned in the flexbasing strategy could make important contributions to the quick response and operational intensity of expeditionary packages of airpower.

EVOLVING TO AN EXPEDITIONARY AEROSPACE FORCE

As the needs of today's strategic environment have become more apparent, the concept of expeditionary aerospace operations has gained strength and momentum. The work to date on the EAF concept has of necessity focused on the organizational aspects of implementation and on accommodating the current continuing demands for forward-deployed forces. However, an expeditionary force must be a rapidly deployable force, not just a rotational force. To achieve this, AEF force packages must have small initial support footprints as well as a large measure of operational and logistical flexibility. One way of achieving such flexibility is the flexbasing strategy that we describe in this report. Such initiatives are a way to keep the process of evolving to a truly expeditionary force moving forward.

AB	Air Base
ABD	Airbase defense
ABL	Airborne laser
ACC	Air Combat Command
ACS	Agile Combat Support
AEF	Aerospace Expeditionary Force
AEW	Aerospace Expeditionary Wing
AFB	Air Force Base
AFI	Air Force Instruction
AFRC	Air Force Reserve Command
AFSOC	Air Force Special Operations Command
AMC	Air Mobility Command
ANG	Air National Guard
AOR	Area of responsibility
APOD	Aerial port of debarkation
ATO	Air tasking order
AWACS	Airborne Warning and Control System

BCAT	Beddown Capability Assessment Tool
BOT	Bombs on target
BSP	Base support plan
C2	Command and control
C2I	Command, control, and intelligence
C4ISR	Command, control communications, computers, intelligence, surveillance, reconnaissance
CALCM	Conventional air-launched cruise missile
CAS	Close air support
CASF	Composite Air Strike Force
CAT	Category
CB	Chemical/biological
CEP	Circular error probable
CHEM/BIO	Chemical/biological
CINC	Commander-in-Chief
COB	Co-located operating base
COLPRO	Collective protection
CONOPs	Concept of operations
CONUS	Continental United States
CSAR	Combat search and rescue
CSL	Core support location
CW	Chemical warfare
DARPA	Defense Advanced Research Projects Agency
DCAPES	Deliberate Crisis Action Planning Execution System

DEW	Distant Early Warning
DOD	Department of Defense
EAF	Expeditionary Aerospace Force
Equiv	Equivalent
FMSE	Fuels management and support equipment
FOL	Follow-on operating requirements
FSL	Forward support location
GMTI	Ground moving-target indicator
GRL	Global Reach Laydown
IOR	Initial operating requirements
JCBAWM	Joint Chemical/Biological Agent Water Monitor
JDAM	Joint Direct Attack Munition
JSTARS	Joint Surveillance and Tracking System
LANTIRN	Low Altitude Navigation and Targeting Infrared for Night
LD/HD	Low density/high demand
LOG	Logistics
LOGCAT	Logistician's Contingency Assessment Tools
MAF	Mobility Air Forces
MDS	Mission design series
MOB	Main operating base
MOG	Maximum on ground
MRE	Meal, Ready to eat
Muni	Munitions

NATO	North Atlantic Treaty Organization
NR	Not relevant
NRO	National Reconnaissance Office
OCA	Offensive counter-air
OPLAN	Operations plan
Optempo	Operations tempo
ORI	Operational Readiness Inspection
OST	Order and ship time
PAA	Primary aircraft authorized
PAC	Patriot advanced capabilty
PACAF	Pacific Air Forces
PGM	Precision-guided munitions
POL	Petroleum oil and lubricants
RF	Response force
RF	Radio frequency
RV	Reentry vehicle
SAC	Strategic Air Command
SAM	Surface-to-air missile
SAR	Side-looking airborne radar
SEAD	Suppression of enemy air defenses
Shel	Shelter
SOF	Special operations force
SSC	Small-scale contingency
STEP	Survey Tool for Employment Planning

Ston	Short ton
Sup	Supplement
SWA	Southwest Asia
TAC	Tactical Air Command
TALCE	Tanker Airlift Control Element
TASS	Tactical Automated Surveillance System
TBM	Tactical ballistic missile
TCF	Tactical combat force
TDS	Theater distribution system
TEMPER	Tent, extendible, modular, personnel
TFA	Toxic free area
TPFDD	Time-phased force and deployment database
TUCHA	Type Unit Data File
UAV	Unmanned air vehicle
UN	United Nations
USAF	United States Air Force
USAFE	United States Air Forces Europe
Veh	Vehicles
WMD	Weapons of mass destruction
WOC	Wing operations center
WRM	War readiness materiel
WWX	Worldwide express

Chapter One

INTRODUCTION: THE EXPEDITIONARY IMPERATIVE

The U.S. Air Force has a long history of adapting itself to meet its country's changing security needs. In the past, as the defense challenges facing the United States have changed, the Air Force has sometimes radically changed the emphasis it places on the various missions it performs and how it bases its forces. Now, with the fundamental changes that have occurred in the global security environment in recent years, the Air Force is again transforming itself to meet new defense challenges. In August 1998, General Michael Ryan, the Air Force Chief of Staff, announced that the Air Force would begin a process of remolding itself into an Expeditionary Aerospace Force (EAF).[1] The nature of this transformation, the challenges it presents, and the associated solutions are the subjects of this report.[2]

In any organization the size of the U.S. Air Force, there are obstacles to significant institutional change, as well as substantial costs that must be paid to implement it. For this reason, it is worthwhile considering whether a transformation to a new concept of operations is necessary. In this report, we will start by briefly describing the un-

---

[1]The Air Force defines "expeditionary" as conducting "global aerospace operations with forces based primarily in the US that will deploy rapidly to begin operations on beddown." Headquarters USAF, *Expeditionary Aerospace Force Factsheet*, June 1999.

[2]In this report, the term Expeditionary Aerospace Force, or EAF, will refer to the current efforts to transform the Air Force as an institution into an expeditionary military force. The term Aerospace Expeditionary Force, or AEF, will refer to each of the ten groupings of forces the Air Force is using to schedule eligibility for overseas deployment. Forces that deploy overseas from these AEFs will be referred to as "AEF force packages," or sometimes as "expeditionary aerospace forces."

derlying rationale for the current Air Force basing and deployment concepts, concepts that have their roots in the early years of the Cold War. We will then consider how the strategic landscape has shifted, the problems this has caused for the Air Force as an institution, and how a more expeditionary concept of conducting operations is intended to address these problems.

ANOTHER GLOBAL STRATEGY, ANOTHER AIR FORCE

The Air Force has a history of transforming itself to meet new strategic challenges. In 1952, the Air Force was forward-based, and had plans to become even more so. As shown in Figure 1.1, the Air Force planned to build 82 overseas bases, many of which were for basing the medium bombers of the Strategic Air Command (SAC).[3] This was the early implementation of the post-war U.S. strategy to contain Soviet expansionism, a strategy that was underwritten by a ring of alliances and bases around the Soviet Union. The Air Force concept of operations relied on a combination of permanent, forward-deployed B-47s on these bases and a mobility plan called Reflex. The plan envisioned the rapid deployment of hundreds more B-47s from the continental United States (CONUS) to overseas bases upon the receipt of strategic warning.[4]

By 1954, this expeditionary, forward-oriented basing approach began to look vulnerable. Improvements in Soviet nuclear delivery capability led Wohlstetter to warn,

> Analysis of the consequences of a Russian A-bomb air attack on the whole of the projected 1956 overseas primary-based system with the projected defenses clearly shows that only small numbers of A-bombs are needed to eliminate the majority of the force . . .[5]

---

[3]In the years immediately following World War II, these consisted chiefly of B-29 bases. By 1954, when the limited speeds of the aircraft made them vulnerable to Soviet defenses, they were being replaced by B-47s. Harkavy (1983), p. 116.

[4]The Reflex deployments are an early example of an "expeditionary" employment of airpower. Each deploying wing was to require the equivalent of 40 C-54 cargo aircraft loads to haul support equipment and personnel to the forward operating location. The bomber crews and aircraft were expected to remain at the forward base for "several days" before generating the first combat sortie. Wohlstetter (1954), p. 4.

[5]Wohlstetter (1954), p. xxvi.

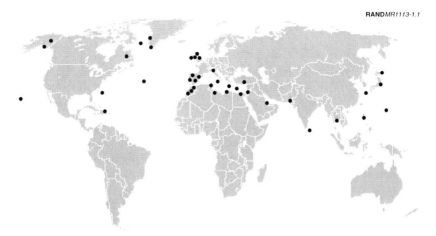

RAND*MR1113-1.1*

SOURCE: Wohlstetter (1954), p. 6. Wohlstetter's map does not portray all 82 of the programmed overseas bases.

Figure 1.1—Programmed Overseas SAC Base Structure in 1952

The warning times available to disperse and launch SAC bomber forces were the critical factor. With the distant early warning (DEW) system of radar sites in place, CONUS bases could expect about two hours of attack warning, whereas overseas bases would get as little as 10 minutes. Wohlstetter recommended that SAC adopt a CONUS-only mode of basing, extending the range of the B-47s with a combination of aerial refueling and overseas transit bases. At the time, SAC was not only forward-deployed but had inherited a World War II CONUS basing structure with most installations located in the south, where the weather allowed more productive aircrew training. Once the decision was made to move to a CONUS-only basing system, a far-reaching adjustment of basing structure and employment concepts took place. A massive base-building campaign was funded to rapidly build SAC bases in the northern tier of states, where transcontinental polar missions could be supported.[6] By the 1960s, the more optimal basing structure, along with the procurement of longer-range B-52s and KC-135s, allowed the Air Force to reach the

---

[6]Futrell (1989), p. 512.

Soviet Union from its bases in CONUS without, for the most part, depending on overseas bases. As the B-47 was phased out, the number of SAC overseas bases was rapidly drawn-down until only one remained, a main operating base (MOB) at Anderson Air Force Base in Guam.

It was because of the nature of the strategic threat described above, and the knowledge that a long-term struggle with the Soviet Union required a sustaining infrastructure, that the Air Force came to be essentially a garrison force. Not only bombers but also tactical fighter forces normally operated from MOBs, with permanent installations located in CONUS, Europe, and the Far East. Even those CONUS-based fighter units that were intended to deploy overseas in the event of war relied on plans to join an elaborate overseas structure of allied collocated operating bases (COBs) at which extensive facilities, support personnel, and equipment were already in place.[7] Over time, this Air Force approach of employing forces from major installations became an enduring part of Air Force culture.

THE COMPOSITE AIR STRIKE FORCE

The Air Force has not always gone "first class." At around the time of the decision to bring the strategic bomber force home to CONUS, there was a countervailing initiative on the part of the Tactical Air Command (TAC)—the Composite Air Strike Force (CASF), which was operational between 1955 and 1973. The CASF concept was intended to provide rapidly deployable CONUS-based tactical forces that could be sent overseas in response to smaller-scale contingencies. Brigadier General Henry Viccellio, the first commander of the 19th Air Force, described the mission of the CASF in this way: "As SAC is the deterrent to major war . . . so will the Composite Air Strike Force be a deterrent to limited war."[8] Many of the issues and challenges being encountered today by the Air Force in its movement toward the EAF were presaged by this early expeditionary concept.

---

[7]Harkavy (1989), p. 82.

[8]Brig Gen Henry P. Viccellio, USAF, "Composite Air Strike Force," *Air University Quarterly Review*, 9:1, Winter 1956–1957, pp. 27–38, as quoted in Futrell (1989), p. 450.

A CASF consisted of a small planning and command element, provided by the 19th Air Force Headquarters, to which regular tactical fighter and fighter-bomber units were attached. In 1958, CASFs were deployed to Incirlik Air Base, Turkey, during the Lebanon crisis, and to Taiwan during the Chinese attacks on the islands of Quemoy and Matsu. Both of these deployments encountered a number of problems that would be familiar today, including trouble with overflight rights, shortages of spare parts and munitions, and a lack of training opportunities while deployed. However, in the Incirlik deployment, 36 F-100s were deployed within 24 hours.[9] The complete CASF, consisting of about 100 F-100s, B-57 tactical bombers, and RF-101 and RB-66 tactical reconnaissance aircraft arrived within four days.[10] Figure 1.2 shows part of the CASF encampment at Incirlik Air Base. Note the tents adjacent to the flightline used for sleeping accommodations, as well as the small size of the nearby munitions

SOURCE: Department of the Air Force (1997), p. 42.

Figure 1.2—CASF Deployment Site, Incirlik AB, Turkey, 1958

---

[9]Futrell (1989), p. 612.
[10]Tilford (1997), p. 112.

storage area. Support for this force consisted of 860 personnel and 202 tons of equipment.[11]

The Air Force did not commit significant research, development, procurement, or training funds to advancing the CASF concept. At the time, the Air Force was focused on its central mission—strategic nuclear deterrence. What is relevant today, however, is the underlying rationale for the CASF, which grew out of the perception that a regional deterrent capability was needed to address communist aggression around the wide periphery of the Eurasian landmass. Instead of maintaining a system of expensive regional MOBs, the Air Force elected to rely on a rapid deployment capability to swing its CONUS-based forces to wherever they might be needed. Today's AEF, conceived to deal with a similar situation of unpredictable and widely dispersed regional crises, is in many ways a modern heir to the CASF.

AN EXPANDED STRATEGIC PERIMETER

Despite the concerns in the 1950s about containing unpredictable Soviet and Chinese incursions into the Eurasian rimland, the central reality of the Cold War for both superpowers became the maintenance of the strategic nuclear deadlock. Many longstanding ethnic, religious, and national conflicts were placed in a kind of stasis—held in check by their implications for the broader superpower struggle. Today, these conflicts have reemerged, along with an array of additional challenges, including "failed states," nonstate terrorism, and international criminal organizations. The United States considers that its security is closely tied to the maintenance of regional stability, partly to foster economic development and the growth of overseas markets. The emergent security challenges, while sometimes threatening U.S. "vital interests," do not threaten the existence of the United States in the same sense that the nuclear struggle with the

---

[11]Tilford (1997), p. 117. A recent AEF deployment consisting of 36 F-16s and F-15s required 1100 personnel and 3200 tons of equipment and munitions. Although the deployed support requirements of combat aircraft have grown since 1958, it should be noted that the CASF support concept consisted of a small replacement spares package, no component repair capability, and few munitions. The crews were not trained for conventional weapons delivery—the weapons of choice were tactical nuclear weapons.

Soviet Union did. So while the challenges may be important ones, the United States will usually have a smaller stake in each outcome than will the opposition, whether that opposition is an aggressive regional hegemon or a drug cartel. At the same time, the security threats have become harder to anticipate and plan for. These developments pose problems for defense planners, and represent what amounts to a much-expanded "strategic perimeter"—not just expanded spatially, but expanded with respect to the type and timing of the threats as well. Like a number of great powers in the past, the United States must meet a series of often unpredictable security challenges around the edge of a hard-to-manage frontier. This is the new security environment in which the Air Force has found itself, one for which its defining Cold War experience had not well prepared it.

THE OLD AIR FORCE MEETS THE NEW DEFENSE ENVIRONMENT

The new environment began to manifest itself immediately after the Gulf War. Since that time, the Air Force has been called on to support an almost nonstop series of crises and lesser contingencies. As shown in Figures 1.3 and 1.4, these have ranged from humanitarian operations in Africa to shows of force in the Middle East and coercive airstrikes in Bosnia. Each of these deployments has represented a substantial Air Force effort. For example, Figure 1.5 shows just one of these contingencies, the PHOENIX SCORPION deployment to the Middle East in November 1997. To deploy the relatively small force shown, many thousands of people and tons of equipment, representing a large support "footprint," were rapidly deployed, then redeployed home again a few months later.

The constant drumbeat of these contingencies during the 1990s has taken a toll on the Air Force, and shows no indications of slackening. Indeed, Figure 1.6 indicates that the trend is toward higher fractions of the Air Force deployed forward from their MOBs on a temporary or rotational basis. This rise is a function of both an *increase* in the numbers of temporarily deployed personnel (approximately 4000 in

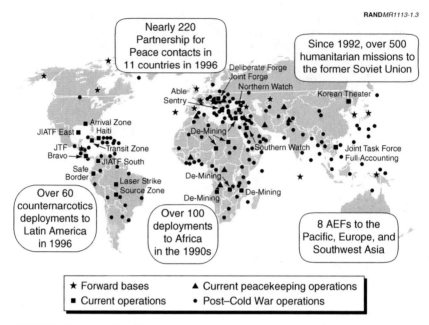

RAND*MR1113-1.3*

Figure 1.3—Global Air Force Operations in the 1990s

1989 versus 17,600 in 1998) and a *decrease* in the pool of deployable personnel (approximately 368,000 in 1989 versus 212,000 in 1998).[12] This represents almost an eightfold increase in the proportion of Air Force personnel deployed from the end of the Cold War in 1989 until today. In fact, the proportion of Air Force personnel deployed temporarily overseas today is nearly that of 1990, when Operation Desert Storm required major overseas deployments.

These trends have resulted in the types of pressures shown below:

---

[12]The pool of deployable personnel was roughly estimated by using the number of personnel in force-providing commands—Air Combat Command (ACC), Air Mobility Command (AMC), Air Force Special Operations Command (AFSOC), United States Air Forces Europe (USAFE), and Pacific Air Forces (PACAF).

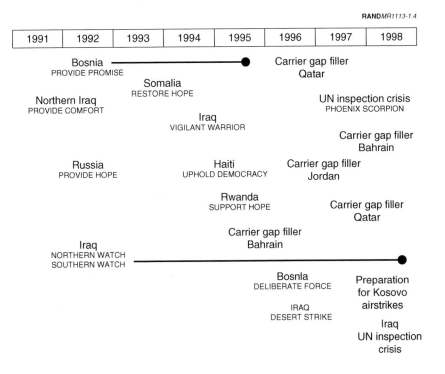

**Figure 1.4—Crises and Contingencies Drive Air Force
Operations Tempo**

- Frequent deployments of specialized units
- Difficulty in maintaining home bases when troops deploy abroad
- Unpredictable assignments
- Lower levels of readiness
- Retention problems.

Many units, called "low density/high demand" (LD/HD), have specialized training or system capabilities that are constantly in demand overseas by regional commanders-in-chief (CINCs).[13]   These units

---

[13]Examples are special forces, rescue, AWACS, and JSTARS units.

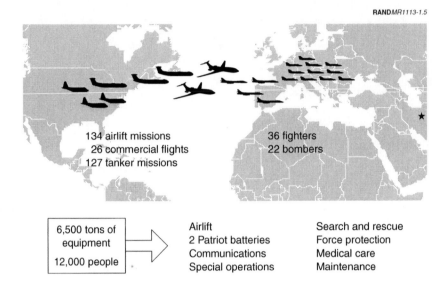

Figure 1.5—PHOENIX SCORPION Deployment Requirements

have had extremely high rates of deployment, often exceeding 180 days per year. In addition, when Air Force units are deployed overseas from their home MOBs, they take with them many of the personnel that are needed to keep those bases operating, such as security police, aircraft maintenance specialists, and civil engineers. This puts a marked strain on the members remaining behind, in some cases requiring 12-hour shifts, seven days per week, for the entire 90- to 120-day period of the deployment. The taskings themselves are by their nature unpredictable, and make it difficult for families to make plans. Finally, training and readiness have suffered, as deployed personnel do not have the opportunity for quality training to maintain their combat skills. The combined effects of these factors have resulted in drastically declining retention rates in many Air Force career fields. Using the pilot force as an example, the Air Force saw a

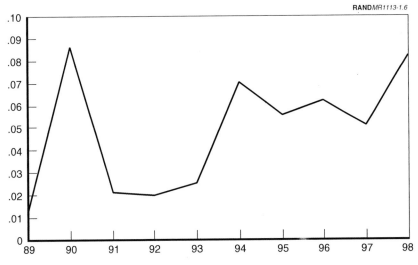

SOURCES: DFI (1997), p. 45; Department of the Air Force (1992), p. 34.

**Figure 1.6—Proportion of USAF Personnel Deployed Overseas on
Temporary Duty at Any One Time**

shortfall of 800 pilots in 1998 and is expecting 1200 in 1999, rising to
2300 by 2002.[14,15]

In 1997, the Quadrennial Defense Review recognized that the crises
and small-scale contingencies (SSCs) that have been causing these
problems for the Air Force were defining characteristics of the new
security environment. It formulated a new defense strategy tailored
to these challenges that emphasizes regional engagement. The strategy has three precepts. The first is to *shape* regional security situations to foster stability and deter aggression. The second is to maintain a capability to rapidly *respond* if deterrence fails. The last is to
prepare now by modernizing our forces.[16] It is to support this new

---

[14]Graham (1998), p. 1.

[15]Correll (1998), p. 4.

[16]Cohen (1997), p. 9.

strategy that the Air Force is again recasting its doctrine and concepts of operation.

THE EXPEDITIONARY AEROSPACE FORCE

The Air Force's answer to the new strategy, the EAF concept, was an outgrowth of several years of thinking and experimentation. The need for a better expeditionary capability became apparent during the VIGILANT WARRIOR deployment to Southwest Asia in October 1994. In response to a movement of Iraqi armor toward Kuwait, the Air Force rapidly deployed more than 200 aircraft to Southwest Asia. The many deployment problems encountered by the Air Force during this contingency indicated that its combat forces needed to be more deployable.[17] In particular, the large support "footprint" mentioned earlier needed to be addressed.

General John Jumper, then commander of the Air Force component of U.S. Central (CENT) Command, responded with the concept of the AEF. The original AEFs were packages of 30 to 36 CONUS-based combat aircraft that were available to the regional CINCs on short notice.[18] From 1995 through 1998, they were deployed six times to Southwest Asia in response to contingencies, as well as to periodically augment theater forces with the ground-attack equivalent of a carrier battle group. Although these deployments went more smoothly and rapidly, they were by no means done on the spur-of-the-moment. Each one followed months of deployment planning, training, aircraft preparation, and movement of support equipment to the pre-identified forward operating locations (FOLs).

If allowed sufficient strategic preparation, these force packages of aerospace power were and still are capable of rapid deployment. This rapid deployment capability, combined with the fact that they were a pre-identified force capability provided to CINCCENT, sometimes allowed the AEF force packages to be left at their CONUS home

---

[17]These problems included the transportation of unneeded and redundant equipment, personnel processing delays, and a lack of information about the mission and destination location.

[18]These AEF force packages generally achieved the stated goal of being able to place "bombs on target" within 48 hours of receipt of a deployment order.

bases until they were needed. This was seen by the Air Force as a way to better manage the operations tempo (optempo) of its personnel.

Another aspect of the original AEF concept was that it served to highlight the garrison structure of the Air Force, since at the time there was little doctrinal guidance for the deployment and employment of expeditionary forces, nor was it clear how they should fit into a joint command structure. Although much uncertainty about expeditionary operations remained, the drawdown in overseas bases along with the continuing need for forces overseas made it clear that the Air Force as a whole needed to become generally more expeditionary. The process culminated in 1998 with an announcement by General Ryan that the service intended to restructure itself into an Expeditionary Aerospace Force, or EAF.

The EAF concept has two primary goals. The first is to provide greater stability and predictability to Air Force personnel by periodically rotating the burden of deployment eligibility around the entire force of airmen. This is intended to address the readiness, training, and retention problems described above. The second goal of the initiative is to enhance the utility of aerospace forces to joint commanders by improving their deployability and tailorability. The Air Force plans to accomplish these goals by establishing ten servicewide AEFs. These AEFs are different from the original 30 to 36 aircraft concept. As shown in Figure 1.7, each AEF will be made up of about 175 aircraft: fighters, bombers, and aircraft performing search and rescue, command and control, reconnaissance, tactical airlift, and aerial refueling. Although the ten AEFs will not be identical to each other, the Air Force intends to compose them so as to possess roughly equivalent combat capabilities, including precision air-to-ground weapons delivery, air superiority, suppression of enemy air defenses (SEAD), and tactical airlift. In addition to the ten AEFs, two wings will be permanently designated as "rapid response" wings to reinforce the capability of the Air Force to react quickly to crises.

RAND*MR1113-1.7*

Forward deployed	Capabilities	On call
18 x F-15C	Air-to-air	6
10 x F-15E	PGM	14
8 x F-16CJ	SEAD	10
12 x A-10 (6 units)	Anti-armor/CAS	14 (ANG)*
3 x E-3	Surveillance/C2	0
3 x HH-60	CSAR	9
8 x C-130 (2 units)	Intratheater	10 (ANG)*
4 x KC-10	Air refueling	2
3 x KC-135 (2 units)	Air refueling	7 (AFRC)*
3 x KC-135 (2 units)	Air refueling	7 (ANG)*
3 x C-21A	Transportation	6
0 x B-52/B-1	CALCM/SA	6
0 x B-2	Stealth	3
0 x F-117	Stealth	6

| 75 | ⟶ | 175 total | ⟵ | 100 |

AFRC = Air Force Reserve Command
ANG = Air National Guard
CALCM/SA = Conventional Air-Launched Cruise Missile/Standoff Attack
CSAR = combat search and rescue
PGM = precision-guided missiles

SOURCE: Ryan (1998), p. 14.

Figure 1.7—Example Aerospace Expeditionary Force Composition

At any given time, two of the ten AEFs and one of the rapid response wings will be eligible to deploy their forces overseas.[19] Each will remain in this status for 90 days before being replaced by another AEF and rapid response wing. This will establish a regular 15-month rotation cycle for the AEFs to enhance the stability of the force. When an AEF enters into its period of deployment eligibility, some of the force will be deployed immediately to support continuing

---

[19]Because AEFs will not all contain the same aircraft, supported CINCs will be asked to request capabilities and not specific aircraft types. It is still uncertain whether the CINCs and their staffs will go along with this restriction, especially if they do not view the weapon systems as equivalent. Also, the CINCs' ability to support aircraft may vary, and may indicate the deployment of specific aircraft types.

overseas commitments.[20] This is intended to support the *shaping* precept of the National Military Strategy, since these forces will be fully engaged and involved in regional security affairs. Remaining behind at their home bases will be a substantial force ready for crises and SSCs. This "on-call" force is intended to be rapidly deployable, to support the *respond* aspect of the defense strategy. The types of contingencies for which this on-call force could be tasked to respond range from humanitarian, to show of force, to early combat against forces invading a friendly country. Although they are envisioned as being rapidly deployable, deployments by AEF force packages will not always need to be fast or early arriving. The timing and phasing of expeditionary aerospace forces into an overseas theater of operations would be at the discretion of the regional CINC.

ORGANIZATION OF THIS REPORT

After announcement of the EAF initiative, the Air Force began planning to implement the concept, with the goal of having the first two AEFs ready for deployment by October 1999. This activity has focused chiefly on the following activities:

- Assigning of specific units to each of the ten AEFs

- Establishing the rotational cycle

- Sourcing of additional support forces to relieve the burden on units assigned to the AEF home bases

- Planning for the deployment of less than squadron-sized numbers of aircraft

- Developing training plans for AEF preparation

- Incorporating expeditionary concepts into institutional culture and training.

The research described in this report does not attempt to duplicate the work in these areas. Instead, it addresses a specific challenge re-

---

[20]As of the time of this writing, such commitments consisted of the NORTHERN and SOUTHERN WATCH no-fly zone enforcement over Iraq, as well as the DENY FLIGHT missions over Bosnia.

lated to successfully executing rapid expeditionary deployments—gaining access to overseas bases. In Chapter Two, we propose a strategy for global aerospace presence. Next, we examine two key enabling aspects of this strategy. Chapter Three addresses the capability to deploy and sustain AEF force packages at locations anywhere throughout a region of instability—wherever access is provided or operations become necessary. However, without a robust defensive capability, enemy threats could deny access to expeditionary forces. In Chapter Four, we discuss the possible threats to deployed aerospace forces, and the defensive capabilities those forces will need to ensure that their deployment options are not reduced.

Chapter Two

FLEXBASING: A STRATEGY FOR GLOBAL AEROSPACE PRESENCE

An overseas presence strategy is a key issue that must be addressed by the Air Force in its evolution to an EAF. During crises and contingencies, operating locations within a region are essential for commanders to have the flexibility and responsiveness they need to manage rapidly unfolding events. In addition, during peacetime, overseas presence provides stability, deterrence, and the capability to shape the security environment favorably. In this chapter, we describe an approach to the base access question that we call *flexbasing*. The essential elements of this strategy are not new. It does, however, bring together in one concept much of the current thinking about overseas presence, base access, new regional support concepts, "bomber islands," and the role of space. We propose that the Air Force take a global view of support for aerospace power projection, and that a global basing strategy will be as important to the EAF concept as was the CONUS-basing of bombers during the Cold War.

BASES IN THE WRONG PLACES

It is not surprising that the basing structure designed to support U.S. strategy during the Cold War is not aligned to meet the unpredictable and varied threats of today. The Air Force's base structure at the end of that experience was shaped by a combination of political, economic, and military forces, resulting in essentially two sets of bases. One set was centered on Europe to support NATO, the other in Japan and South Korea to back up the bilateral security agreements with those two countries. In 1981, there were 41 of these bases. Today, this number has been drawn down to 13, concentrating on Western Europe and Northeast Asia. These remaining bases are shown in Figure 2.1, overlaid with circles or ellipses showing major areas of instability. Also shown is a sampling of sites where the Air Force has had to conduct operations on a temporary or rotational basis since 1991. This illustrates in plain terms the "expanded strategic perimeter" described earlier, and that the old basing model of conducting operations directly from MOBs is not supporting Air Force operations as well as in the past. Building new MOBs in the many areas of instability around the world is neither affordable nor politically realistic. A new approach to overseas presence is needed that can provide true (not "virtual") presence, and that can address the demands of the widened U.S. security perimeter.

RAND*MR1113-2.1*

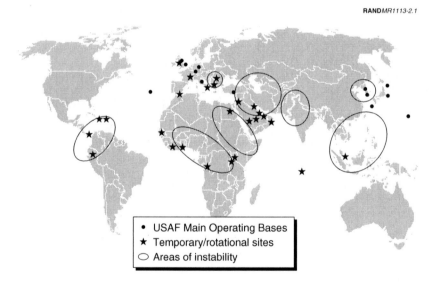

Figure 2.1—USAF Overseas MOBs and Regional Instability, 1999

MEETING THE CHALLENGE OF AN EXTENDED SECURITY PERIMETER

In the past, the great powers that successfully faced this type of security challenge used the following types of capabilities:[1]

- Economy of force
- Global awareness (intelligence)
- Advanced command, control, and communications
- Rapid mobility
- Advanced logistics
- Cultivation of regional partners.

Economy of force was achieved by centrally marshaling forces in secure areas. Good intelligence about the frontier area, combined with rapid communications and decisionmaking capabilities allowed threats to be identified and prioritized. Rapid mobility and logistics capabilities allowed forces to be moved swiftly to where they were needed and sustained for as long as necessary. Finally, regional friends and allies often allowed the threats to be handled locally, without military action from the center.

It is remarkable how closely these capabilities—needed for security in the next century—correspond with the capabilities and core competencies of today's Air Force. Richard Kugler has written extensively about the demands that the broadened strategic perimeter place on the United States, and says of the Air Force,

> USAF forces appear well situated to play roles of growing importance in the coming years. Both agile and well-armed, USAF forces, more than any of the other services, can project power over long distances quickly when bases and infrastructure are available to receive them.[2]

---

[1]For relevant historical examples, see Luttwak (1976), and Kennedy (1983).

[2]Kugler (1998), p. 116.

The key is infrastructure. Kugler recommends that as part of the implementation of the U.S. engagement strategy there be a high-priority effort to establish a network of reception facilities in the "outlying areas" of strategic instability. These often low-profile facilities would provide prepositioned equipment and spare parts to enable the rapid projection of aerospace power.

> By developing a large network of these bases, USAF forces could quickly deploy to many new locations, thus providing an early reaction capability in the critical period before U.S. ground and naval forces can arrive in strength.[3]

Kugler calls these facilities deployment operating bases. We call them forward support locations (FSLs) and forward operating locations (FOLs); they represent just a part of the flexbasing strategy. Another part is a hedging approach to access—that is, multiple low-cost deployment sites identified throughout a theater, spread over different countries and geographic areas.

FLEXBASING AND ASSURED ACCESS

The concepts underlying flexbasing are not new. In addition to Kugler, the Air Force Scientific Advisory Board in its 1997 study of expeditionary operations spoke of the need to establish "regional contingency centers" at which support equipment and spares could be positioned.[4]   General John Jumper, who fathered the idea of expeditionary aerospace force packages when he was commander of Air Force forces in Southwest Asia, addressed the access issue by establishing five possible deployment locations in the region, in five countries. In a related concept, others have talked of regional "bomber islands," at which long-range aircraft could be positioned for possible employment throughout a theater. What is new about flexbasing is the realization that the combination of all these ideas represents a useful approach to the access issue. Advanced logistics and mobility support concepts, combined with a robust mix of long-

---

[3]Kugler (1998), p. 133.
[4]Fuchs (1997), p. H-32.

and short-range combat systems, space systems, and new approaches to planning, converge to solve the access issue.

Flexbasing replaces the concerns about achieving an elusive assured access around the world with a robust *capability* to go swiftly and easily into whatever base, airport, or remote airfield becomes available. Few, if any, sovereign countries are likely to give before-the-fact, blanket permissions for combat operations to be conducted from their territories. However, in general, if there were no friendly countries in a region whose security was being threatened, it would be unlikely that the United States would have reason to deploy forces. If friendly countries were being intimidated into denying access, or doubted U.S. resolve, the Air Force would retain the capability to operate from a distance until closer operating locations became available.

THE ELEMENTS OF FLEXBASING

Although complex cost and operational tradeoffs remain on the details of the flexbasing concept, our analysis indicates that the concept has essentially six elements. The actions shown in Figure 2.2 would be necessary to implement the flexbasing strategy for global presence.

Establish a Global System of Tiered Locations

The three types of locations in the flexbasing approach are core support locations (CSLs), FSLs, and FOLs. Briefly, core bases are represented by the CONUS and overseas MOBs operated by the Air Force, FSLs are sites for storing materiel to facilitate deployments throughout a particular region, and FOLs are possible deployment sites in a region. For each type of base, the hedging strategy is applied to ensure that there are multiple options to accomplish the Air Force mission in each region—each identified and anticipated. A possible structure of CSLs, FSLs, and FOLs is shown in Figure 2.3. MOBs (CSLs) on sovereign U.S. territory are clearly the most desirable in terms of access; the remaining overseas MOBs are spread over a number of countries. The situation is best in Europe, where the MOBs are located in a range of allied countries—the United

RAND*MR1113-2.2*

- Establish a global system of tiered locations: core, FSLs, and FOLs

- Maintain a robust mix of long- and short-range weapon systems

- Develop earth orbit as a forward support location

- Advocate global presence strategy as a joint goal, supporting and supported by *shaping*

- Design and establish a global logistics/mobility support system

- Incorporate full-spectrum force protection into basing strategy

Figure 2.2—Elements of Flexbasing

RAND*MR1113-2.3*

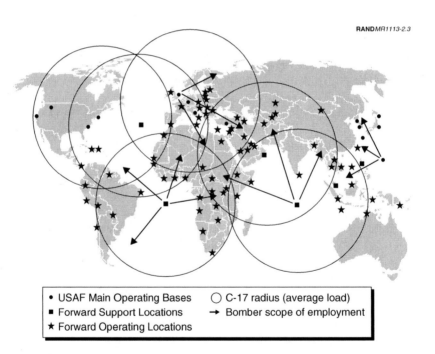

• USAF Main Operating Bases	○ C-17 radius (average load)
■ Forward Support Locations	→ Bomber scope of employment
★ Forward Operating Locations	

Figure 2.3—Notional Flexbasing Concept

Kingdom, Germany, Italy, and Turkey. The situation is more problematic in East Asia, where MOBs exist only in Japan, South Korea, and Guam. In addition, the forces in Japan and South Korea are focused on the deterrence of North Korea. The projection of aerospace power in this region may seem troublesome, but the Air Force should reconsider the advantages of Guam as a CSL for support of the flexbasing strategy. It is well positioned with respect to force protection, and provides an ideal location for the employment of long-range bombers throughout Southeast Asia. Alternatively, Guam could be developed as an FSL, which would allow it to be used for periodic (not continuous) deployments by forces from the forward-deploying parts of the CONUS-based AEFs. This would demonstrate U.S. commitment to the security of the region and would enhance expeditionary training. It would also allow smaller demonstration and training visits to FOLs throughout the region, thus actively supporting the shaping strategy without taxing PACAF forces. The Air Force should consider the buildup of Guam as a key support location for the projection of aerospace power in Southeast Asia.

The general functions served by the three types of locations in support of the flexbasing strategy are shown in Figure 2.4. The categorization is based on the investment in infrastructure and deployable support equipment at the site. Core base functions will overlap those performed by the FSLs, thereby providing a long distance backup in case of delays in accessing an FSL in a region.[5]

The FSLs, an important aspect of the flexbasing concept, provide regional storage sites for support infrastructure, as well as locations at which to beddown bombers, tankers, and other enabling assets (e.g., AWACS, JSTARS, Airborne Laser, and theater airlift). During peacetime, these locations could provide a low-profile presence, possibly involving only warehouses tended by contractors with periodic inspections by Air Force personnel. Given enough deployment time, all of the supporting equipment and munitions positioned at FSLs

---

[5]Access to the bases in the support system will vary, and it may well be that certain FSLs (e.g., Guam, Diego Garcia) may provide greater access than some core bases (e.g., Kadena). The overall regional access capability must be considered. Certain core bases may need to be backed-up with equipment and munitions stored at FSLs.

RAND*MR1113-2.4*

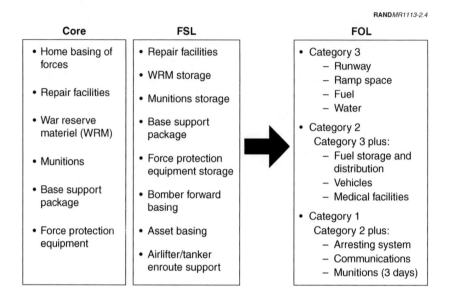

Core	FSL	FOL
• Home basing of forces • Repair facilities • War reserve materiel (WRM) • Munitions • Base support package • Force protection equipment	• Repair facilities • WRM storage • Munitions storage • Base support package • Force protection equipment storage • Bomber forward basing • Asset basing • Airlifter/tanker enroute support	• Category 3 – Runway – Ramp space – Fuel – Water • Category 2 Category 3 plus: – Fuel storage and distribution – Vehicles – Medical facilities • Category 1 Category 2 plus: – Arresting system – Communications – Munitions (3 days)

Figure 2.4—Flexbase Functions

could come from the core bases. However, it is the regional positioning of this equipment at FSLs that allows the rapid deployment and employment of combat aircraft within a matter of hours or days. The FSL capability to support the deployment and sustainment of AEF force packages is discussed in the next chapter.

The FOLs are stocked with prepositioned equipment and munitions according to the level of responsiveness desired. Category 1 FOLs are the most responsive, having the capability to support combat operations within 48 hours of the initial deployment order. To achieve such responsiveness, most of the munitions, vehicles, fuel storage systems, and other supporting equipment must be in place before the forces arrive. Therefore, Category 1 FOLs represent a relatively expensive capability, and would be set up only in regions of vital U.S. interest. Category 2 and Category 3 FOLs are not as responsive and cost substantially less. Category 3 FOLs, in fact, would require only the minimum of in-place support, at (probably) little or no cost to the United States. The keys to moving swiftly into these types of locations will be the regional positioning of equipment at the FSLs and the prior accomplishment of the necessary surveys and Base Support Plans (BSPs). Although the development and updating of a large number of such plans would represent a substantial effort, the payoff would be enormous. In the next chapter, we will describe some automated tools in development to support this requirement.

Maintain a Robust Mix of Long- and Short-Range Combat Capabilities

The Air Force could address the basing issue, as well as the problem of force protection, by relying almost exclusively on long-range weapon systems—bombers and cruise missiles. There are, however, some difficulties with this idea. The Long Range Airpower Panel, in a 1998 congressionally mandated study, concluded that bombers by themselves could not achieve the sortie rates needed by a CINC during high-intensity operations.[6] In addition, for nonstealthy B-52s and B-1s to operate over hostile territory, local air superiority is needed, as well as suppression of enemy air defenses (SEAD). Fighter aircraft currently accomplish these missions, and these aircraft would need to be deployed forward.[7] Finally, in a rapidly changing situation, mobile targets may need to be hit quickly. Today's cruise missiles are effective only against fixed targets, although this could change in the future.

Bombers are, however, an essential expeditionary capability and are important in the flexbasing strategy. Potential host countries often seek evidence of U.S. commitment to "go the distance" in a military operation before providing access. If they observe a continuing series of bomber strikes, potential coalition partners are less likely to be intimidated by a local aggressor and more likely to approve U.S. access to forward bases. Similarly, in the case of punitive strikes or show of force operations, from which friendly regional states may wish to distance themselves, the Air Force would be able to quickly deploy long-range systems to pre-stocked FSLs and launch the op-

---

[6]Welch (1998).

[7]We note that sea-based fighter aircraft could also provide enabling capabilities to support long-range bomber strikes. Such joint solutions should be considered and balanced against the cost of providing support for the deployment of land-based fighters.

eration with a minimum of additional deployed assets. In essence, the flexbasing strategy rests on the concept of having multiple basing and employment options. A mix of long- and short-range weapon systems is a vital aspect of achieving such flexibility.

Develop Earth Orbit as a Forward Support Location

The Air Force should think of space not as a medium to be traversed but as a forward location to support expeditionary operations. Earth orbit, although constrained by Kepler's laws, is an omnipresent forward base located only 200 miles from any area of operations. Space also affords excellent access, and space operations represent a comparative advantage for the United States. Some of the functions that could feasibly be supported from space are:[8]

- Rapidly launched "mini-Milstars" for protected AEF communications

- Rapidly launched or electro-optical satellites

- Space-based lasers for SEAD

- Space-delivered kinetic-kill reentry vehicles for SEAD

- Space-based surface-to-air missile (SAM) radar jamming for SEAD

- Space-based ground moving-target indicator/side-looking airborne radar (GMTI/SAR) for AWACS/JSTARS sensing.

Much of the difficulty in rapidly deploying and employing AEF force packages relates to the array of "enablers" that must also go along to allow successful combat operations. The types of systems listed above could provide many of these functions, such as a communications surge capability, space-based sensing, and SEAD. A high-leverage force enhancement would be the capability to provide AWACS- and JSTARS-like sensing and tracking from space. One program that holds promise in providing these capabilities is the Discoverer II program, which is jointly managed by the Air Force, the Defense

---

[8]Some of this material is taken from Daniel Gonzales and colleagues, "Implications of reusable space systems for Air Force space operations," RAND, internal draft, 1999.

Advanced Research Projects Agency (DARPA), and the National Reconnaissance Office (NRO). This program has the objective of developing and launching two satellites by 2004 to demonstrate the feasibility of providing stereoscopic AWACS- and JSTAR-like sensing from space. In addition, the satellites are planned to provide digital terrain elevation data (needed for low-level mission planning). All data would be provided directly to deployed expeditionary forces via downlinks. Analysis indicates that a constellation of 24 of these satellites could provide instantaneous and near-continuous surveillance of any place on the earth's surface or in the atmosphere.

Moving more of the Air Force mission into space does present some considerable challenges, however. Agreements such as the Space Treaty and the Anti-Ballistic Missile Treaty could constrain weapons-related missions from orbit. However, just being able to accomplish many of the sensing and reconnaissance missions from space would lower the support footprint of deploying AEF packages considerably. Another obstacle is the considerable cost of developing, fielding, and maintaining these systems. These costs need to be balanced against the cost and operational effectiveness of similar atmospheric platforms.

Still, with some enabling functions performed from earth orbit, substantially fewer personnel and aircraft would need to be deployed forward. Expeditionary aerospace forces could become more agile and light, and the deployment burden on "low density/high demand" assets such as AWACS could be resolved. In addition, without the need for in-theater/in-atmosphere supporting platforms, bombers could possibly operate with more independence, enhancing the long-range application of aerospace power from core bases and FSLs.[9]

Advocate a Global Presence Strategy

Any strategy for the global projection of U.S. combat power must be a joint and even national strategy. For example, the positioning of theater-assigned equipment at FSLs and the collection of data on

---

[9]Project AIR FORCE's "Extended Range Airpower Study" is examining options for enhancing the long-range application of airpower.

possible FOLs would need to be endorsed and instituted by the regional CINCs. Agreements for establishing FSLs would require negotiations by the State Department. Nevertheless, it is also true that many aspects of the EAF concept itself will need buy-in and endorsement by the Joint Staff and the regional CINCs, especially with respect to the integration of AEF force packages into operations plans (OPLANs) and other contingency plans that are developed by these players. The Air Force could pursue a global basing policy as an initiative at the joint level, as it pursues other initiatives related to implementing the EAF concept.

Clearly, any overseas military presence policy would need to become a corollary of our overall military strategy. We recommend that such a policy be presented to the joint community as part of the EAF initiative, as advantageous for all the services. FSLs would not be reserved for Air Force use only. All services could use them for locating repair facilities or for positioning support equipment. For the CINCs, the strategy provides better access and would allow quicker response during crises.

Another attractive aspect of flexbasing is that it would be a tangible objective of the shaping aspect of our military strategy. Shaping activities such as military-to-military contacts and training with coalition partners are chronically underfunded, partly because it is difficult to demonstrate the benefits of these activities. Additionally, engagement activities such as these are today conducted at the initiative of the CINCs, without overall funding, policies, or objectives established by the Department of Defense.[10] If one aspect of shaping were defined as the pursuit of a global military presence policy, it would be easier to perceive the benefits. Figure 2.5 presents some of the possible benefits of shaping, as applied to the flexbasing strategy. With measures of effectiveness such as these, a case could be made for better funding of these activities if the results in a particular theater were not up to expectations.

---

[10]Davis (1997), p. 37.

RAND*MR1113-2.5*

- Number of FSLs established and their locations

- Number of FOLs visited by forward-deployed AEF force packages

- Completeness and currency of FOL databases

- Types of military/humanitarian missions that can be supported by basing structure

Figure 2.5—Measurable Benefits from Shaping

Design and Implement a Global Agile Combat Support/ Mobility System

We have already alluded to many of the logistical aspects of the flexbasing strategy. Innovative logistics and mobility concepts will be the *sine qua non* of a global presence strategy for the Air Force. We recommend that attention be given to the design and implementation of an integrated logistics and mobility support system for expeditionary aerospace forces. Additionally, we recommend that this be done on a global basis, not region-by-region. Such a system will be discussed in the next chapter.

Implement Full-Spectrum Force Protection

Another key enabler of an expeditionary basing policy is a force protection capability that addresses the full spectrum of possible threats. Without highly effective and deployable force protection, AEF force packages would not be able to take advantage of the full array of available FOLs. Indeed, they might not be able to move forward at all, defeating the basing strategy. This will be addressed in Chapter Four.

In the near term, we believe that the last two aspects of the flexbasing strategy are the keys to its implementation. Flexibility, the key to the strategy, is provided by a powerful logistics capability to deploy to wherever expeditionary aerospace forces need to go and by not being precluded from going there by enemy threats.

Chapter Three

ENABLING THE STRATEGY: A GLOBAL LOGISTICS/ MOBILITY SUPPORT SYSTEM

Our strategy for global aerospace presence rests on the capability to rapidly deploy forces to a large number of locations with varying characteristics and on having the operational flexibility to employ effectively from those locations. Some locations may be distant from the fight, and long-range weapon systems will be used. Some may be quite close, allowing the responsiveness and intensity of shorter-range weapon systems to be brought to bear, while raising the importance of force protection as a key enabler. As in times past, however, it is superior logistics and mobility capabilities that make possible the defense of an extended strategic perimeter. In this chapter, we highlight the major characteristics of a global logistics and mobility system to support the expeditionary strategy. We also describe the analytic process that the Air Force can use to determine the details of such a system, and provide results from the application of that process.

THE SUPPORT CHALLENGE

Perhaps the greatest challenge the Air Force faces in becoming more expeditionary is overcoming the traditionally heavy nature of its support processes and equipment. Having operated chiefly from MOBs for most of its history, the Air Force has had little need to make items such as avionics test equipment, bomb loaders, and communications gear as light and transportable as possible. For example, the deployment of intermediate-level avionics maintenance for 24 F-15s requires up to three C-17s. Even "deployable" equipment is quite heavy. The deployment of shelters to support a bare-base operation for a typical AEF strike package requires 20 C-17s.[1] The challenge involves more than just the transportability of equipment. It also involves finding new, more expeditionary ways of doing business by reconsidering the levels of initial support and infrastructure needed by deploying forces, beginning sustainment operations immediately, or conducting some support functions such as parts maintenance at FSLs and core bases. New deployment processes and practices have the greatest potential for near-term improvements to AEF force package deployability. Examining and reengineering processes such as maintenance concepts and early beddown requirements will be an important part of making the Air Force more expeditionary.

As a point of departure for our examination of Air Force support processes and equipment, we examined the deployment of an AEF force package, the 4th Aerospace Expeditionary Wing (AEW), to the Persian Gulf State of Qatar in 1997.[2] The logistical footprint associated with that deployment is shown in Figure 3.1. The rapid deployment[3] consisted of only about 20 airlifter missions, representing only 788 tons out of a total of almost 3200 tons of equipment and materiel that were needed to support operations at the forward location. The balance of the requirement was already in place when the forces arrived. In addition, months of planning, specific to the wing and its known destination, were required before the deployment.

This deployment was a significant waypoint on the Air Force's course to the EAF concept, and it highlighted two of our early findings. First, we found that with current logistics processes and equipment, substantial amounts of prepositioned equipment and supplies are a necessity if the ambitious deployment goals of the Air Force, such as 48 hours to "bombs on target," are to be achieved. Whereas new

---

[1]We assume a 30-aircraft AEF package and use of HARVEST FALCON base-support packages to support 1100 people at the forward location. The AEF Battlelab at Mountain Home AFB, Idaho, is examining the feasibility of a new base-support package called HARVEST PHOENIX that would substantially reduce the initial transportation requirement.

[2]This was a deployment of 30 combat aircraft to fill a scheduled "carrier gap" requirement. The package consisted of 12 F-15Es, 12 F-16CGs, and 6 F-16CJs.

[3]The wing achieved the goal of generating combat sorties at the forward location after receiving 24 hours of strategic warning and 48 hours to actually deploy.

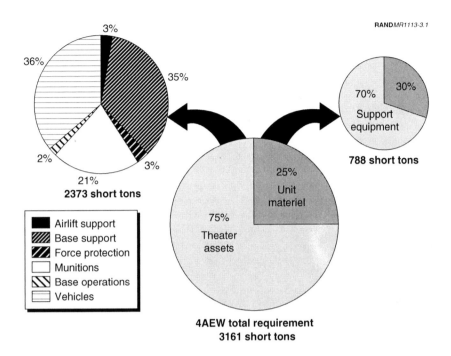

Figure 3.1—Breakdown of Support for 4th AEW

technologies will improve this situation in the mid to long term, implementing the EAF over the next few years will require many in-place, prepositioned resources.

Reducing the overall deployment footprint of deploying EAF forces will be an evolutionary process, involving the procurement over time of lighter and more-deployable support equipment, as well as more-supportable weapon systems. However, we found that the greatest near-term improvements in EAF deployability could be achieved by changing support practices and policies. This led to our second finding, that the best opportunities for improving the situation are in the *strategic* decisions about the logistics and mobility processes that are involved with deploying and sustaining forces. The 4th AEW deployment focused on streamlining the deployment execution (see the processes shown in Figure 3.2). The figure illustrates the current ACC standard for deployment (72 hours of strategic warning, 24

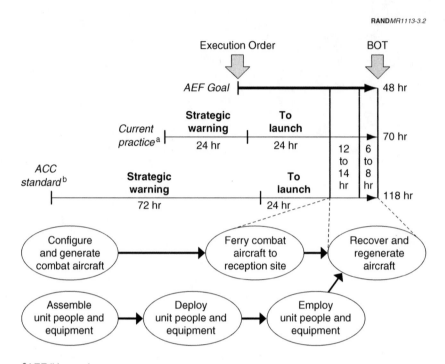

RAND*MR1113-3.2*

[a]AEF IV experience.

[b]Phase I Operational Readiness Inspection (ORI) for 24 PAA (primary aircraft authorized) units (AFI 90–201, ACC SUP 1, January 1, 1996).

Figure 3.2—Deployment Timelines

hours to start deployment, followed by another 18–24 hours to arrive in theater, regenerate the aircraft, and begin to launch strikes. The ovals list the tasks to be executed when strategic warning is given. The 4th AEW made substantial improvement on that timeline.

We found that the biggest payoffs will be achieved by examining the strategic decisions that must be made long before the deployment takes place. Figure 3.3 illustrates the relationship of strategic decisions to the execution decisions shown in Figure 3.2. Of the strategic decisions shown, our research focused primarily on those regarding forward infrastructure—which Kugler (1998) pointed to as critical to

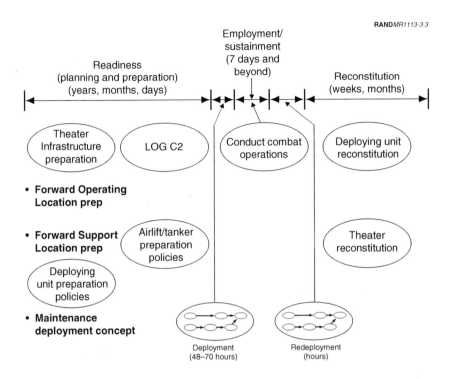

Figure 3.3—Strategic Deployment Decisions

the projection of aerospace power. We found it to be an important element of an overseas support structure for the Air Force.

ELEMENTS OF A GLOBAL LOGISTICS/MOBILITY SYSTEM

Decisions about what to preposition, and where, form the basis of infrastructure preparation. There are tradeoffs to be made between a number of competing objectives, including responsiveness, cost, footprint, risk, and flexibility. Prepositioning everything at the forward location improves responsiveness, but it also reduces flexibility, adds political and military risk, and incurs a substantial cost if a number of such bases are to be prepared. Bringing support from CONUS or an in-theater location increases flexibility and reduces risk, but results in longer timelines and requires increased airlift.

Considering these tradeoffs, there are essentially five elements of a logistics and mobility system to support expeditionary aerospace forces. The first three—FOLs, FSLs, and CSLs—have already been introduced as important aspects of the flexbasing strategy. Here we will consider on their logistics and mobility aspects.

- **Forward operating locations.** As indicated earlier, there are three categories of FOLs, with each category requiring different amounts of equipment to be brought in to make the base ready for operations. Each therefore has different timelines and transportation requirements. A key decision about theater infrastructure is deciding how many FOLs of each type the Air Force needs in a critical area.

- **Forward support locations.** FSLs are regional support facilities outside of CONUS with high assurance of access but not located in a crisis area. FSLs can be joint depots for U.S. WRM storage, for repair of selected avionics or engines, a transportation hub, or a combination of these. They could be manned permanently by U.S. military, by host nation personnel, or simply be a warehouse operation until activated. The exact capability of an FSL will be determined by the forces it will support and by the risks and costs of positioning specific capabilities at its location. FSLs will have an enhanced potential for using local military or contractor facilities to support regionally engaged AEF force packages.

- **Core support locations.** CSLs are MOBs located both in CONUS and overseas. They are the home bases for Air Force forces, and provide the full range of operations support. Some core bases will back up the FSLs, providing repair capabilities and deployable supplies and equipment.

- **An air mobility network** will connect the FOLs, FSLs, and CSLs, including en route tanker support. If AEF force packages are to deploy leanly, rapid and assured transportation links are essential. FSLs themselves will likely be transportation hubs and beddown sites for air mobility forces.

- **A logistics C2 system** will coordinate the entire support structure, organize transport and support activities, and allow the system to react swiftly to rapidly changing circumstances.

Strategic decisions about the global support system for the expeditionary forces will require choices about the roles that each of these elements should play, considering the security challenges in each region. One choice will involve how many Category 1 FOLs are needed to support a rapid response to aggression in a given region. Another could concern the types of supplies to be placed at each FSL—ranging from munitions to humanitarian supplies. Yet another could be a decision about where component repair should take place. Our research has provided a framework for analyzing these questions, along with some initial answers.

ANALYTIC FRAMEWORK FOR STRATEGIC PLANNING

Process Models for Evaluating Support Options

The core of our analytic framework for strategic logistics/mobility planning is a series of models of critical support processes that can calculate equipment, supplies, and personnel required to support operations at an FOL. Because support requirements are a direct function of mission requirements, the models must be employment-driven; that is, they start from the operational scenario with estimates of types and numbers of aircraft, sortie rates, types of weapons, and so forth. Once the support requirements are computed, we need to evaluate options for satisfying those requirements—for example, prepositioning the equipment, deploying it from CONUS, or deploying it from regional support locations. The evaluation considers several dimensions, such as spin-up time (the time required for the deployed force to be ready to conduct operations from its deployed location), footprint (the amount of airlift capacity the deployment requires), peacetime costs (both investment and recurring), flexibility, and risks (both military and political). Figure 3.4 depicts the framework. This process is repeated for each of the resources or commodities needed at the FOL. For our analysis, we developed models to estimate the requirements for munitions, POL support, unit maintenance equipment, vehicles, and shelters. These requirements account for the bulk of the support needed at an FOL.

RAND*MR1113-3.4*

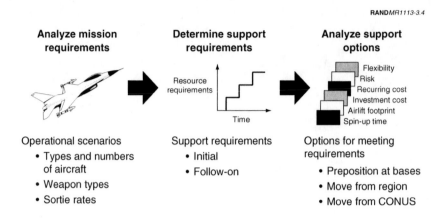

Analyze mission requirements

Operational scenarios
- Types and numbers of aircraft
- Weapon types
- Sortie rates

Determine support requirements

Resource requirements

Time

Support requirements
- Initial
- Follow-on

Analyze support options

Flexibility
Risk
Recurring cost
Investment cost
Airlift footprint
Spin-up time

Options for meeting requirements
- Preposition at bases
- Move from region
- Move from CONUS

Figure 3.4—Employment-Driven Analytic Framework

The primary advantage of employment-driven models for making strategic support decisions is that they allow us to deal with the pervasive uncertainty of expeditionary operations. The models can be run for a variety of mission requirements selected by operators, allowing examination of support performance for different types of missions (humanitarian, evacuation, small-scale interdiction, etc.), the effects of different weapon mixes for the same mission (e.g., new, light munitions), and other potential modifications to the theater environment.

To use the support models in this manner, the models must run quickly and estimate requirements at a level of detail (numbers of personnel, pallets, and large pieces of equipment such as fuel trucks, bomb loaders, cranes, etc.) appropriate for the strategic decision. At the same time, they must contain enough detail so that major changes to the process can be reflected and evaluated in terms of their effects on different metrics. For example, one insight gained from our research is that the requirements for some support processes can be divided into Initial Operating Requirements (IOR)—the equipment, people, and supplies needed to begin operations—and Follow-on Operating Requirements (FOR) needed for sustainment. Being able to distinguish these in the model provides a more flexible set of options for providing the necessary support.

The next step compares an option's capability and cost with those of other options. This allows the tradeoffs to be observed between options involving the movement of resources from CSLs, from FSLs, or prepositioning them at the FOLs from which the aircraft will fly. Mobility requirements enter the process here as well. For example, prepositioning equipment at FOLs reduces mobility requirements and spin-up time, but at higher costs (to preposition sets of resources at a number of FOLs) and possibly greater risks (access to the equipment would be subject to political or military interference). Positioning resources at an FSL or CSL is less expensive but extends the spin-up time and assumes the availability of substantial amounts of airlift. Figure 3.5 shows a sample tradeoff between cost and spin-up time for munitions support in a scenario where heavy bombs would be used for ground attack. The bars show the cost for each of the munitions storage options, and the lines show an optimistic to pessimistic

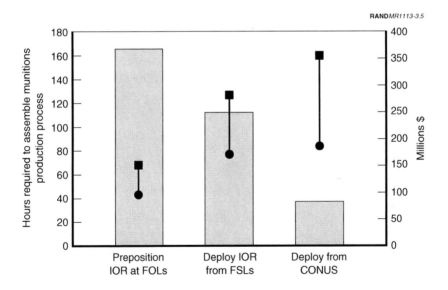

Figure 3.5—Tradeoff Between Cost and Spin-Up Time

range of responsiveness.[4]   Note that substantial levels of possibly costly prepositioning are necessary to achieve the highest levels of responsiveness.

Integrating Models for Design of Overall Support Concepts

Models of individual support processes can yield important insights into support processes for expeditionary operations. However, for strategic planning of a logistics/mobility system, we need to integrate the outputs of models of different processes and consider mixes of options. Support concepts could include a mix of prepositioning some materiel, deploying other materiel from FSLs, and deploying still other from CONUS. To choose among all the options for each resource group, we developed a prototype mixed-integer optimization model. The use of optimization techniques, which have a long history of application to logistics planning and analysis, was a way to identify feasible least-cost support concepts. This automated tool selected one or more support options in each of the commodity areas, using the criteria of responsiveness and cost. Taken together, these options represented a possible support concept for expeditionary aerospace operations that could then be examined more closely to consider additional issues, such as the operational flexibility of the concept and its transportation feasibility. Figure 3.6 lists the main characteristics of the model.

When we applied this model to the positioning of munitions, fuel, vehicles, and shelter for a single theater—Southwest Asia, the results were as shown in Figure 3.7. We chose 48-, 96-, and 144-hour deployment timelines as benchmarks for three types of FOLs. The Category I FOL, with the most in-place equipment, provides the most responsive capability. If less responsiveness is allowable, more supplies and equipment can be provided from FSLs and CONUS, which provides planners with better flexibility with regard to possible operating locations. Note that, in general, little support can be provided from CONUS unless even longer deployment timelines are

---

[4]The "Deploy from CONUS" option is considered more variable because the greater distances imply greater risk of delay arising from maintenance problems, diplomatic clearances, and the like. The airlift flow in the analysis was constrained by a maximum on the ground (MOG) of two aircraft at the FOL.

RAND*MR1113-3.6*

- Selects a candidate mix of support options
- Minimizes peacetime cost, subject to:
  - Wartime sortie generation requirements
  - Deployment timeline requirements
  - Limitations of air mobility system
  - Robustness requirements
- Models transportation network
  - FOL, FSL, and CSL locations
  - Distances/travel times
  - Aircraft, ships, and trucks
  - MOG limitations
- Multiple theaters are possible
  - Sharing of logistics locations between theaters

Figure 3.6—Integrating Model Characteristics

accepted. Another important result of this analysis was that in every case, the regionally located FSL was an essential contributor in least-cost solutions. This would seem to support the FOL-FSL-CSL aspect of the flexbasing concept from both a cost and deployability perspective.

As the Air Force extends its analysis of support structures beyond single theaters of operation, the complexity of the tradeoffs involved will make the application of automated techniques such as those illustrated here even more essential. The complex interactions between the region-specific security challenges, mutually supporting theaters, cost, and required levels of responsiveness will create many possible support structures. In addition, as new and more deployable equipment is being considered, or new policies and procedures are formulated, their effects on the overall cost and deployability of the EAF concept will be difficult to judge without an integrated and automated analytic framework.[5]

---

[5]For a more thorough discussion of integrated strategic planning for an ACS/mobility system, see Tripp et al. (1999).

RAND*MR1113-3.7*

Timeline	FOL	FSL	CSL
48 hours (Category 1)	Bombs (IOR) Fuel FMSE Shelter Vehicles	Missiles (IOR & FOR) Bombs (FOR) Repair: avionics and engines	Unit equipment
96 hours (Category 2)	Bombs (IOR) Fuel Shelter Vehicles	Bombs (FOR), FMSE Repair: avionics and engines	Unit equipment Missiles (IOR & FOR)
144 hours (Category 3)	Fuel	Bombs (IOR & FOR) Repair: avionics and engines Shelter Vehicles	Unit equipment Missiles (IOR & FOR) FMSE

NOTES: Deployment times and distances are based on Southwest Asia.
FMSE = fuels management and support equipment.

Figure 3.7—Least-Cost Resource Positioning to Meet Timeline Criteria

SYSTEM DEPLOYMENT PERFORMANCE[6]

Employment Scenario and Metrics

As described earlier, our analytic method uses employment scenarios to derive logistics requirements. In the analysis described here, we addressed a scenario that places heavy demands on those commodities (munitions, POL support, unit maintenance equipment, vehicles, and shelters) that account for most of the support footprint. The scenario is illustrative of the type of questions that can be answered by our analytic framework. Other missions, weapons, sortie rates, etc. could also be examined to evaluate the robustness of any proposed support concept.

---

[6]The discussion in this subsection is taken from unpublished work by Lionel Galway et al.

The scenario elements that determine the requirements for the major commodities are the number of aircraft and their types [mission design series (MDS)], their sortie rates, their missions (which determine the munitions they carry), and their munitions expenditure rates. The key outputs of the models are the people, equipment, and consumables. In this example analysis, we will focus on the cost and weight of the equipment and consumable items. We converted the weight into airlift requirements by using standard planning factors.

The scenario illustrated here is based on the deployment experience in Southwest Asia (SWA), since it has been in this theater that the concept has been most tested. Although the aircraft, missions, and sortie rates are taken from the Air Force component of Central Command (CENTAF) experience, we believe that the experience is useful in addressing the support needs of AEF force packages more generally. The basic AEF force package in the analyses below consists of

- 12 F-15Cs for air superiority,

- 12 F-15Es for ground attack with GBU-10s, and

- 12 F-16CJs for SEAD missions.

In our baseline scenarios, these aircraft execute 80 sorties per day (rates of 2.3, 2.3, and 2.0 sorties per day, respectively).[7] We consider only the equipment and material required to conduct the first seven days of operations.[8]

In comparing the performance of different infrastructure components both individually and in different configurations, we use five metrics: deployment timeline, deployment footprint (equipment and people), peacetime cost, flexibility, and risk. Our analytic method provides quantitative treatment of the first three, which will be described in more detail below. Flexibility and risk were ad-

---

[7]This is a demanding scenario, and some Air Force planners have questioned whether such a small force could sustain this optempo for seven days.

[8]Seven days has emerged as a canonical planning parameter for the initial operation. Clearly, if combat operations are initiated and extended beyond seven days, daily resupply will be a necessity.

dressed subjectively, although ongoing research is considering ways to more systematically evaluate these factors. Figure 3.8 displays the

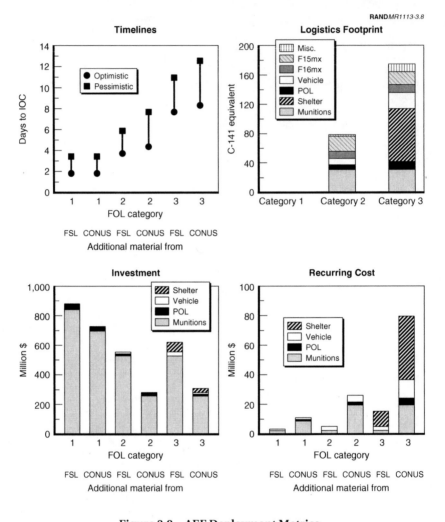

Figure 3.8—AEF Deployment Metrics

metrics estimated with the employment-driven process models for six support Concepts of Operations (CONOPs). The metrics are displayed together to facilitate comparisons.[9]

Timelines to Deploy

For a Category 1 FOL, the optimistic time to set up the base is just under two days, even though most equipment is prepositioned. This result is primarily driven by the time to deploy personnel from CONUS and to set up munitions and fuel-storage facilities.

For the rest of the options, the times are primarily driven by the MOG and by the assumption of C-141s as the transport aircraft. The difference in timelines between CONUS and an FSL is minimal because the bottleneck is in unloading.[10] For Category 3 bases, the primary time driver is unloading the bulky HARVEST FALCON package. Setting this package up requires four to six days with a dedicated 150-man crew, in a temperate climate.[11]

In summary, meeting the 48-hour timeline will be virtually impossible with current processes and equipment unless most equipment is prepositioned. Even then the timeline is extremely tight.

Deployment Footprint

We define the deployment footprint as the amount of equipment and material that must be moved to the FOL for operations to commence.[12] The footprint is derived from the model outputs: the model computes the equipment and vehicles needed for each commodity, and then converts this to airlift requirements using standard planning factors for each selected aircraft (raw short tons could be

---

[9]A feature of the process models, called a TradeMaster, facilitates these comparisons.

[10]This assumes that the tanker airbridge, which can add time, has already been deployed.

[11]It is current Air Force practice to set up complete HARVEST FALCON sets before declaring an initial combat capability. This could change as more austere base-opening packages are proposed and approved.

[12]As indicated in the previous footnote, the size of the deployment footprint can change with changes in support policy.

used as well).[13] The upper right-hand panel of Figure 3.8 shows the initial airlift requirements for the three categories of FOL (i.e., the amount of airlift required to get the base operating).

Peacetime Cost Estimates

Although transportation and material costs are of secondary importance when a crisis looms, fiscal concerns require that part of the evaluation of any set of options include the peacetime costs of setting up and operating the system. These are shown in Figure 3.8 as investment and recurring costs. To estimate the costs, we assumed that there were two theaters of operation covered by the system, with an FSL in each theater. To implement the hedging strategy for base access discussed in Chapter Two, we assumed that there were five FOLs in each theater. We also assumed that the system needed the capability to support two simultaneous AEF force package deployments per year.

As expected, providing for five Category 1 FOLs per region is expensive, and munitions are by far the greatest cost (although recall that only the munitions IOC is prepositioned at each base). Drawing materiel back to the FSLs decreases the cost, increases flexibility, and (may) decrease risk because each FSL requires only two sets of equipment. However, airlift requirements are increased.[14]

The recurring costs have two components—transportation costs for exercising the system with force package deployments twice a year, and the storage and maintenance costs for the equipment stored at the various locations. The lower right-hand panel of Figure 3.8 shows our estimate of the recurring costs for the base configurations we are examining. These recurring costs show a different pattern

---

[13]The actual computations are a hybrid. For most equipment, we compute the weight in short tons and divide by the capacity of the aircraft used for airlift planning purposes. For some bulky equipment, we also use area taken up to correct the computation, or, in some cases, the pallet positions required by the shipment. The measures are usually quite close.

[14]The investment costs do not include costs for building new FSLs. These could be considerable, but are highly dependent on the nature of the relationship with the host country. In addition, some of the costs we counted could be sunk, meaning they have already been paid. However, the costs include those associated with the periodic maintenance and inspection of the equipment stored at FSLs.

from the investment costs—now the Category 3 bases supported from CONUS are relatively expensive to operate, primarily because of the large costs of transporting munitions and the HARVEST FALCON sets twice a year for exercises.

Looking at Figure 3.8 as a whole, we can see that Category 1 FOLs give the fastest response but at a high investment cost. As one might expect, Category 2 FOLs have a longer response time but at a lower investment cost. In general, stockpiling at FOLs has higher investment costs than stockpiling in CONUS, but it has lower recurring costs. These costs provide useful insights into the sources of cost for the flexbasing concept. We believe that these observations are robust across a wide range of scenarios, and that they will need to be taken into account in a broader analysis of the structure of the global logistics/mobility support system for expeditionary aerospace operations.

SYSTEM SUSTAINMENT PERFORMANCE[15]

It seems clear that a global network of FOLs, FSLs, and CSLs will be essential for rapid deployments for intensive combat operations. We also find that such a network is required to *sustain* expeditionary forces. FSLs, in particular, will play an important role in sustainment, as can be seen by examining the tradeoffs between transportation time and the requirements of such sustainment processes as aircraft and munitions maintenance.

Figure 3.9 shows some of these tradeoffs. The vertical axis represents the fraction of shipments from CONUS to SWA that can be delivered by the day indicated on the horizontal axis. The left-most curve shows the distribution of expected resupply times for small items (e.g., 150 lb or less) that could be shipped via commercial express carriers. This distribution includes the entire resupply time, including the time from requisition submission to receipt of the item by the customer, and has a mean of about four days (including weekends, holidays, and pickup days). The distribution was generated by using optimistic times for each related process, and by assuming the processes are perfectly coordinated (no delays resulting from weather,

---

[15]The discussion in this subsection is taken from unpublished research by Tripp et al.

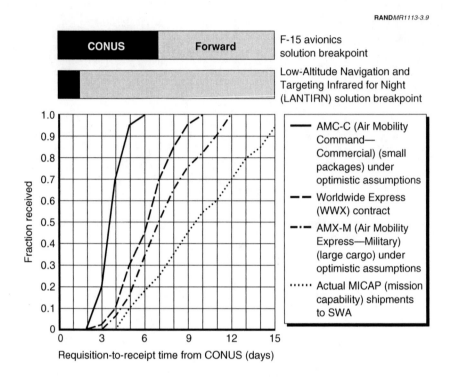

Figure 3.9—Supply Times and Support Breakpoint Solutions

mechanical problems, or enemy actions). The curve is therefore a "process optimum."

The second curve shows the expected distribution of Worldwide Express (WWX) deliveries in a peacetime environment. WWX is a Department of Defense (DoD) contract with commercial express carriers to move small items within CONUS and from CONUS to the rest of the world. The contract has specific in-transit delivery times for shipments between specific locations. For instance, most in-transit times to sites within SWA are about three days, although this time excludes the day of pickup and weekends. With these delays included, the delivery times shown are greater than our optimistic assumptions in the previous curve. The third curve shows the distribution of resupply times for AMX-M, the system used for large cargo in wartime. These delivery times are longer than both the commercial delivery and WWX options. The fourth curve represents the delivery times experienced by an actual AEF deployment to SWA.

We combined this transportation information with data from studies we conducted on two combat aircraft sustainment processes. For component repair of LANTIRN (Low-Altitude Navigation and Targeting Infrared for Night) pods and F-15 avionics components, we calculated the cost breakpoints for (1) locating repair facilities at a CONUS-located CSL or (2) positioning the facilities forward at a location such as an FSL.[16] The results are shown at the top of Figure 3.9. For F-15 avionics, the cost breakpoint occurs at seven days. That is, if delivery time from CONUS is reliably less than seven days, it makes more sense to perform the maintenance in CONUS. If it takes more than seven days, it would make more sense to perform the maintenance regionally, at an FSL. For LANTIRN pods, the breakpoint is lower, about two days, because there are fewer parts available to fill the delivery pipeline.

The curves show that whereas F-15 avionics could be supported from CONUS if the transportation times reached our commercial "best-case" estimate of six days to deliver 100 percent of shipments, the LANTIRN would still be better supported from a forward-positioned regional maintenance capability. However, the real-world and contractual experience shown in the other curves suggests that transportation performance will not come close to the best-case estimate. In addition, it is unlikely that the Air Force would rely solely on commercial package carriers to resupply a unit conducting combat operations. These considerations make forward maintenance preferred for F-15 avionics as well.

The overall peacetime cost of the sustainment system is an important concern. Centralizing certain maintenance functions at FSLs may help contain costs by consolidating assets, reducing deployments for technical personnel (who could be assigned to FSLs during AEF on-call periods), use of host-nation facilities, and possibly sharing costs with allies. Further, considerable infrastructure, including

---

[16]The third possibility—performing the maintenance at each FOL in the theater—is not shown because the FOL option was clearly the most costly, as a result of the expense of deploying multiple sets of test equipment to a large number of FOLs.

buildings and large stockpiles of war reserve materiel, may already be available in areas such as Europe.

Our analysis has indicated that many support functions such as component repair and engine maintenance will be provided cost-effectively from regional locations. Regional FSLs will play an important role in both the sustainment of deployed AEF force packages, as well as in their deployment.

TYING THE SYSTEM TOGETHER: MOBILITY AND DEPLOYMENT SYSTEMS

We have indicated that planning for the logistics/mobility support system must be global in scope. The need for this global perspective is perhaps no better demonstrated than by the inherently global nature of the transportation network that will support it. We have proposed a worldwide system of FOLs, FSLs, and CSLs with characteristics tailored to the defense challenges of the various regions in which they are located. Very much in keeping with air mobility doctrine going back to World War II, these sites will be connected by air links that will regularly go into and through a number of CINCs' areas of responsibility (AORs), providing mutual support between them. The transportation services needed to support the flexbasing system with channel services, special airlift missions, aerial refueling, and theater airlift must be centrally planned to provide a global EAF deployment capability.

The implications of flexbasing for the air mobility system are yet to be fully understood and are the subject of ongoing research. However, it is clear to us that the chief issues related to mobility support for expeditionary aerospace forces do not hinge on the availability of sufficient numbers of airlift aircraft per se. The Air Force already has a large fleet of airlifters, and movement constraints such as en route and destination infrastructure almost always come into play before the number of available airlifters runs out. The best opportunities for improving air mobility support to the expeditionary operations lie instead in improving mobility and deployment processes. These processes were developed during the Cold War, when plans and requirements were much more stable and predictable. More dynamic and flexible processes are needed to support today's expeditionary operations. We next discuss the air mobility processes needed to support our flexbasing strategy during peacetime, and then during contingencies.

The System During Peacetime

The roles of the mobility air forces (MAF) in supporting the EAF concept and the logistics/mobility support system during peacetime fall into two categories:[17] those missions needed to maintain the system over time and MAF participation in the forward deployments of AEF force packages.

To maintain the logistics/mobility system during peacetime, the air mobility system will:

- Deploy and redeploy the forward-based AEF force packages every 90 days

- Support the surveillance and maintenance of equipment stored at FOLs and FSLs

- Deploy and redeploy avionics technicians, munitions maintenance specialists, etc. to FSLs

- Discretely build up or decrease regional capabilities as security challenges change

- Test wartime routes used by assured resupply missions; gather resupply statistics for planning wartime sustainment operations.

It appears that the chief peacetime support of the mobility system to expeditionary aerospace forces will lie in enabling the regular deployment and redeployment of the forward-based AEF force packages. Although the true effect of these requirements on MAF Optempo is still being evaluated, preliminary analysis has indicated that the effect should not greatly exceed the current demands being placed on the mobility system. These periodic movements closely mirror the current support given to operations such as NORTHERN WATCH, SOUTHERN WATCH, and DENY FLIGHT.

---

[17]The MAF is a term that includes all mobility forces in the Air Force, including those assigned to AMC, PACAF, USAFE, AFSOC, ANG, and AFRC.

In addition to the force movements, however, air mobility will play a critical role in maintaining the system by providing channel and contract air carrier services. These flights will be a regular U.S. presence at the FOLs and FSLs, and move personnel and equipment to maintain system integrity. The flights will allow the quiet reapportionment of capabilities around the system as circumstances change. Finally, the periodic exercise of the assured resupply routes that will be used during wartime will allow the collection of essential statistics on order-and-ship time (OST) throughout the global system, enabling better logistics planning.

The MAF will not simply maintain the system. It will also deploy its own forces as part of the AEF force packages. These forces can:

- Regularly forward-deploy theater airlift and tanker forces to unstable regions

- Support "shaping" activities

- Train with coalition partners

- Visit for humanitarian and goodwill reasons

- Visit potential FOLs

- Deploy forces to build up en route and theater systems during crises.

As they have in numerous deployments over the past decade, MAF forces will deploy forward with the combat aircraft to provide aerial refueling and theater airlift support to employment operations. In addition, regular short-term visits by air mobility forces to forward areas will advance the shaping aspect of U.S. military strategy. Visits by C-17s, C-130s, or KC-135s are less sensitive than visits by combat aircraft and perhaps present more opportunities for training with potential coalition partners. For example, many other countries around the world fly C-130s, making airdrop training a natural way to interact with foreign militaries. In addition, because of their lower profile, MAF forces could more easily visit potential FOLs within a region, testing air traffic control services, instrument approaches, terrain clearances, fuel availability, ramp space, and other critical parameters that need to be known to implement the flexbasing strategy. Finally, with a regular and nonthreatening air mobility presence throughout the system, it will be easier to quietly deploy theater air-lifters, mission support forces, and tankers during periods of heightened tensions. This could allow a "leg up" on building the airbridge to support rapid expeditionary deployments.

The System During Contingencies

The air mobility system has been conducting deployments of U.S. forces during crises and contingencies for many years. Most of the mobility processes that the Air Force has developed are clearly applicable to the rapid deployment of AEF force packages. In addition, many "expeditionary" concepts, such as the quick projection of infrastructure and bare-base operations, were presaged by such MAF practices as the Global Reach Laydown of support forces and the beddown of theater airlift at austere locations. There are, however, a number of processes that are holdovers from a time when the deployment system was designed solely to execute major operations plans (OPLANs). These OPLANs, and their associated deployment requirements [time-phased force and deployment databases (TPFDDs)], are years in the planning, and there was little perceived need to rapidly assemble coherent deployment requirements or to react to unforeseen events. In what follows, we recommend changes to deployment processes to better support rapid expeditionary deployments.

Quicker Assessment of Deployment Requirements. The EAF concept assumes rapid deployment of aerospace forces that are tailored to the needs of a CINC in a particular crisis. However, today's deployment system was designed to simply execute OPLANs and transport their deliberately planned TPFDDs. There is little capability to rapidly tailor deployment requirements to the needs of a specific crisis. Efficient automated tools are needed to quickly identify and integrate the deployment requirements of AEF force packages as a crisis unfolds.

The chief determinant of deployment requirements is the level of support that is available at the intended FOL. If the deployment is to a Category 3 bare-base FOL, for example, the requirements will be much greater than if it is to a Category 1 FOL. Also, the flexbasing strategy assumes the capability to deploy on short notice to any of a possibly large number of sites within a region. As indicated in

Chapter Two, this will require the collection and storage of a great deal of information, and the capability to disseminate that information to deployment planners during a crisis.

There is a pair of automated tools that holds promise to provide these capabilities. The Survey Tool for Employment Planning (STEP) and the Beddown Capability Assessment Tool (BCAT) are part of a suite of applications called the Logistician's Contingency Assessment Tools (LOGCAT). STEP allows survey teams using laptop computers to efficiently collect the information needed on potential FOLs, including digital photos and video. Base Support Plans (BSPs) can then be developed and made available during crises to logistics planners at all levels. If AEF force packages are to be deployed quickly into available locations, it is critical for logistics planners to have access to pertinent high-quality data.

BCAT is a system that draws on both the forward-location data provided by STEP and on employment requirements from the Air Tasking Order (ATO). Figure 3.10 shows a BCAT interface that uses the data to compare the sortie-generation requirements of the commander's employment plan with the overall logistical capability of a base to generate the sorties. This information will be useful for quickly assessing AEF deployment requirements and tailoring them to the needs of a particular contingency. STEP and BCAT are examples of the types of systems that can help implement expeditionary operations and the flexbasing strategy.

Ability to Rapidly Develop TPFDDs. During a crisis, requirements can change rapidly. A CINC may cancel the deployment of one unit in favor of another as the tactical situation unfolds. The beddown locations of deploying units can change at the last minute. To accommodate the dynamic and fluid situations that expeditionary forces are intended to address, deployment requirements must be quickly combined and put into operationally valid TPFDDs. A system in development that is designed to do this is the Deliberate Crisis Action Planning Execution System, or DCAPES. DCAPES will draw on an array of information systems, including STEP and BCAT, to allow Air Force planners to rapidly assemble the TPFDD for deploying AEF force packages. The ability to rapidly assess, tailor, and assemble deployment requirements should be considered a fundamental ex

RAND*MR1113-3.10*

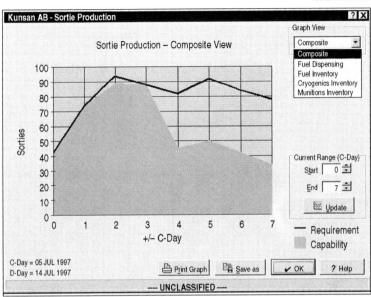

SOURCE: Hunter (1998a), p. 22.

Figure 3.10—BCAT Shows Support Capability Versus Operational Requirements

peditionary capability. Some of the functions that DCAPES will perform are

- TPFDD editing
- TPFDD sourcing
- Manpower tailoring
- Personnel monitoring
- Manpower/personnel feasibility analysis
- Personnel rotations
- Unit type availability/apportionment
- Unit type development/TUCHA (type unit data file) updates
- Logistics planning

Use of "Playbooks." Today most deployment planning takes place in the deliberate planning process in support of the major OPLANs. When contingencies pop up that do not have associated OPLANs, the deployment planning must take place quickly, in the so-called "crisis action planning process." In the past, this has led to confusion and, in some cases, poor plans. There is a multitude of important details that must be taken into account in planning a deployment, and in the midst of a crisis there is often not enough time. In addition, proper attention may not be paid to the flexibility and robustness of the plan, causing it to fall apart at the first unforeseen event.

If expeditionary aerospace deployments are to be responsive and reliable, there needs to be a level of preplanning that is flexible to the situation but addresses many of the details that could be overlooked when time is short. One initiative already undertaken by the MAF to provide more flexible and responsive plans is the idea of developing contingency "playbooks." As the EAF concept is currently envisioned, AEF force package deployments will draw from forces originating from bases across the country. These deployments will be complex, and rapid deployments will need to be carefully choreographed. A "playbook" is a mini-deployment plan that can be tailored to the current on-call AEFs. The level of detail can vary, however, depending on the number of situations the plan is intended to cover. For the MAF to support the expeditionary operations and the flexbasing strategy, less-detailed playbooks should be developed to cover various types of contingencies in each theater of operation. These could include robust and flexible concepts of operations (CONOPs), surveys of the available theater resources (e.g., MOG, fuel, billeting), and what FOLs could become available.

Enhance Capability to "Lean Forward." The deployment of Global Reach Laydown (GRL) packages [i.e., mission support teams and Tanker Airlift Control Elements (TALCEs)] and the setup of tanker airbridge operations are necessary precursors to the rapid deployment of AEF force packages. Currently, the need to position this critical infrastructure has motivated the provision in MAF playbooks of a 24-hour period of strategic warning prior to the deployment exe-

cute order.[18] If this infrastructure positioning can be given a head start in deploying, the overall responsiveness of expeditionary aerospace forces will be enhanced. With a "warm" en route structure designed around the system of FOLs, FSLs, and CSLs, and with play-book plans, it would be easier to deploy certain elements of infrastructure overseas prior to the execute order. It should be standard procedure to move tankers, theater airlift, and mission support forces out into the system during times of heightened tensions. In addition, as discussed earlier, periodic rotational deployments to key sites by tankers and theater airlifters, along with combined training exercises with regional partners, could increase the likelihood of having air mobility forces where and when they are needed. MAF forces would conduct these deployments during their AEF deployment eligibility periods.

Another way to enhance the capability to "lean forward" is to simply decrease the need for such preparatory activity. The amount of positioning at on-load sites by airlifters could be minimized. AEF force package deployments require the movement from many locations of much non-unit equipment and personnel to provide support functions, such as force protection, combat communications, space support teams, Patriot batteries, and chemical warfare defense. Each of these requires a positioning movement by airlifters, which is accomplished during the strategic warning period prior to actual deployment. These movements could be minimized by co-locating many of the support functions with the airlift, that is, at the airlift bases themselves.

We have outlined the essential characteristics of a global logistics/mobility support system for expeditionary aerospace operations and described how this system could support the rapid deployment of AEF force packages overseas, as well as how it could sustain these forces. We have also discussed the role of the mobility and deployment systems that will tie the system together. In the next chapter, we address force protection for deployed forces.

---

[18]Most of this pre-execution positioning is done within CONUS, with the overseas GRL deployment waiting until the execute order.

Chapter Four

FORCE PROTECTION FOR GLOBAL AEROSPACE PRESENCE

Even with a powerful logistics and mobility system that provides maximum flexibility with respect to basing options, enemy threats could still preclude expeditionary forces from exploiting those options. Regional allies can make FOLs available, even ones with substantial amounts of prepositioned U.S. equipment, and both the locations and the equipment could be effectively denied to AEF force packages by credible enemy threats. Force protection is essentially partnered with logistics and mobility capabilities in enabling the flexbasing strategy. Together, they allow expeditionary aerospace forces to deploy to where they must to conduct their mission. Although there are many factors that could delay or deny access to forward bases during a contingency, threats from the enemy or from terrorist groups should be minimized by a robust force protection program.

Since the Khobar Towers bombing in Saudi Arabia in June 1996, the Air Force has taken significant steps to reduce the risk to its deployed personnel and equipment. These steps have included the establishment of the Security Forces Center at Lackland Air Force Base, Texas, along with the "standing up" of the 820th Security Forces Group, whose mission is to rapidly deploy to provide security at overseas Air Force operating locations. These organizations are chiefly focused on ground-based threats to Air Force FOLs. Other Air Force organizations are addressing additonal types of threats. The civil engineering community is concerned with chemical and biological threats, and the Information Warfare Center has been created to develop defenses against threats to information. Because the defense of FOLs is a vital aspect of an expeditionary basing strategy, we pulled together information from these separate efforts to build a complete picture of the force protection challenge. Our analysis surveyed the possible ground, air, chemical/biological (CB), and information threats to deployed expeditionary aerospace forces, and examined possible airbase defense (ABD) in light of effectiveness and effect on AEF force package deployability. In this chapter we will briefly describe the range of force protection challenges that we surveyed, followed by recommendations on enhancing expeditionary force protection capabilities.[1]

CHARACTERIZING THE THREATS AND RESPONSES

In surveying the threats and responses to deployed AEF force packages, we categorized the threats as shown in Table 4.1. Because our purpose in addressing force protection issues was chiefly to gauge the effect on the flexbasing strategy, we devised the threat categories

Table 4.1

Threat/Response Matrix

	Low Intensity ABD Threat Level I	Medium Intensity ABD Threat Level II	High Intensity ABD Threat Level III
Ground	Irregular/terrorist Penetrating threats	Special operations forces (SOF)/sabotage Standoff threats	Regular infantry assaults
Air	"Plane bomb"	Enemy offensive counter-air (OCA)	Theater ballistic missiles (TBM)
CB	Contamination of food and water	Localized chemical warfare (CW) agent contamination	Basewide CW agent contamination
Information	Physical attack on information systems	Disruption of reachback capability	Radio frequency weapons

---

[1]For a more thorough discussion of threats to deployed AEF force packages and specific recommended responses, see Killingsworth (1999).

to be comprehensive but also distinct with respect to their deployment footprint. For example, the responses to conventional ground threats involve different types of equipment and defense strategies than the responses to air threats. We did not separately address the nuclear threat to forward bases, believing that the only effective countermeasure to such attacks would be to prevent their delivery from the ground or air. On the other hand, although CB attacks would be delivered by ground- or air-based weapons, post-attack countermeasures with deployment implications are possible, leading us to specify a separate category. We included information threats, since the use of automated systems and rapid communications will be important to deployed expeditionary operations. In each of these threat categories, we specified low-, medium-, and high-intensity levels, believing that the intensity of the threat would affect the type of response.

We used this threat/response matrix to evaluate how force protection requirements could affect expeditionary basing. For each cell in the matrix, we examined the nature of the threat and the range of current and near-term responses.

THREATS ON THE GROUND

To define the three levels of ground-threat intensity shown in Table 4.1, we relied on the current airbase defense levels used by the Air Force security forces. These threat levels are defined in Table 4.2.

Low-Intensity Ground Threats

The low-intensity ground threat corresponds to Airbase Defense Level I, and includes threats from irregular forces and terrorists. Attempts to penetrate the perimeter of the base were also included in this category. Highly mobile security forces with robust detection capabilities are the best way to counter threats such as these.[2] The

---

[2]Shlapak (1995), p. 66.

Table 4.2[3]

Airbase Defense Threat Levels

Threat Level	Examples	Response
I	Agents, saboteurs, sympathizers, terrorists	Unit, base, and base cluster self-defense measures
II	Small tactical units, unconventional warfare forces, guerrillas	Self-defense measures and response force (RF) with supporting fire
III	Large tactical force operations, including airborne, heliborne, and major air operations	May require timely commitment of a tactical combat force (TCF)

Air Force is improving its detection capabilities by procuring the Tactical Automated Surveillance System (TASS), which allows the continuous surveillance of areas of approach as well as the point monitoring of key facilities and assets. For mobility, "up-armored" HMMWVs (High-Mobility Multipurpose Wheeled Vehicles) have been fully funded and are in the process of being deployed. The 820th Security Forces Group can deploy a force with these capabilities, having approximately 73 personnel plus weapons and vehicles.[4] The unit can be deployed with approximately 1-1/2 C-5 airlifters.

Although the Air Force capability to meet these threats is much improved, the adequacy of these forces will be situationally dependent. For example, fields of view around the base perimeter could be restricted, or local security forces may not be cooperative. Careful planning and site selection will be necessary.

Medium-Intensity Ground Threats

At the medium level, the ground threat is composed of military special forces and mortars or rockets that can strike from outside the FOL perimeter. These types of attacks are hard to prevent or preempt, and pose a serious threat to forward-deployed AEF forces.

---

[3]Department of the Air Force (1996), p. x.

[4]Buckingham (1998).

Current countermeasures include close coordination with host nation forces, patrols of the standoff footprint area, and maintaining a wide, clear field of view around the perimeter.[5] In addition, passive defense measures can be taken, such as constructing concrete and sandbag bunkers and revetments for shelter during attacks. Considered together, these measures are effective, but it is the strong sense of the force protection community that more needs to be done to counter this type of threat. Improvements to expeditionary aerospace capabilities to meet this threat should include tactical unmanned aerial vehicles (UAVs) for better surveillance, and counterbattery and countersniper systems for better firepower against standoff threats. The additional deployment burden imposed by these new systems on deployment requirements is unknown but could be substantial if deployability is not considered during their procurement.

High-Intensity Ground Threats

Although joint rear-area doctrine requires that airbases use their own resources to defend against Level I and II threats, the Level III threats represented by regular brigade-sized infantry forces are planned to be countered with a joint tactical combat force.[6] The defensive measures against this threat would be provided at the theater level from outside of the FOL. Although it is unlikely that an AEF force package would be deployed so far forward that regular forces could threaten it, if such a situation developed the force would probably be redeployed to a location farther in the rear.

THREATS FROM THE AIR

Low-Intensity Air Threats

Small, slow-moving UAVs or suicide "plane bombs" pose an as-yet undemonstrated but potentially serious threat to deployed expeditionary forces. Inexpensive, Global Positioning System (GPS)-guided

---

[5]The deployment of AC-130 gunships might also provide wide-area surveillance and substantial firepower, but these assets are under the control of the CINC, and could easily be re-prioritized to perform other missions.

[6]Department of Defense (1996), p. I-7.

UAVs armed with cluster munitions or chemical agents would be hard to detect and could greatly degrade the sortie-generation capability of an FOL. They would be particularly hard to detect because of their slow speed—existing U.S. air defense systems are designed to filter out slow-moving returns to reduce ground clutter.[7]

The current countermeasures for this threat rely on an active intelligence operation and close cooperation with the security forces and air traffic control system of the host nation. Stillion and Orletsky propose that small machine-gun teams would be effective against slow-moving UAVs, as would man-portable Stinger missiles.[8] The teams would need to be equipped with infrared and optical sensors with which to detect and evaluate unknown aircraft, and be in direct contact with local air traffic control authorities. New concepts of operation, training, and equipment are needed to address slow-moving threats from the air.

Medium-Intensity Air Threats

In this category we placed the threat posed by the enemy air force—enemy OCA operations. The current Air Force response to air threats against its bases is to establish and maintain air superiority in the area of operations. If commanders anticipated that achieving air superiority would be difficult or take time, it is likely that expeditionary operations would begin from a distance, possibly moving forward as circumstances permitted.

High-Intensity Air Threats

The high-intensity threat from the air is posed by TBMs and cruise missiles. These threats could seriously disrupt or attenuate an air campaign. However, the current SCUD-B and SCUD-C types of ballistic missiles commonly in the inventories of rogue states are not accurate, and do not pose a military threat as much as they pose a terror threat. To achieve an 80 percent probability of twice cratering a

---

[7]For a thorough discussion of this type of threat to Air Force forward bases, see Stillion and Orletsky (1999).

[8]Stillion and Orletsky, p. 41.

runway, thereby closing it to operations, an opponent would need to launch more than 40 of these types of missiles.[9] However, arming the missiles with cluster bombs or with weapons of mass destruction (WMD) would substantially reduce the number of missiles required, as would the use of GPS guidance.

As for cruise missiles, no country but the United States has used these in combat, chiefly because of the technical challenges associated with developing accurate guidance systems. The cruise missiles currently fielded by potential adversaries, such as the Chinese *Silkworm*, are highly inaccurate. However, by taking advantage of GPS, these weapons could be provided with low-cost guidance accurate to less than 100 meters.[10] Armed with cluster munitions, they could pose a deadly threat to EAF forward operations.

To address these threats in the near term, the Patriot PAC-3 terminal defense system that is in the process of being deployed will mitigate the current TBM threat.[11] In addition, the Patriot is considered to be effective against the generation of cruise missiles in the hands of our potential adversaries today.[12] The Patriot system, however, presents a substantial deployment challenge. Two recent deployments to Southwest Asia with Patriot support have provided a consistent record, as shown in Table 4.3.

Table 4.3

Patriot Battery Deployment Requirements

Requirement	AEF V	Phoenix Scorpion
Cargo (stons)	270.3	281.0
Passengers	90	92

SOURCE: Conner (1997).

---

[9]M. Eisenstein, in unpublished 1995 research, assumes a 300-meter circular error probable (CEP).

[10]Gormley (1998), p. 97.

[11]*Ballistic Missile Defense Organization, Patriot Advanced Capability PAC-3,* Fact Sheet AQ-99-04, Washington, D.C., February 1999.

[12]Gormley (1998), p. 103.

Deployment represents a requirement of approximately eight C-141s and two C-5s in each case, and would take approximately 2-1/2 days to transport and set up. In the meantime, other necessary combat support equipment could be delayed because of ramp space and aerial port limitations.

The Air Force should be seeking a force protection capability against TBMs and cruise missiles that is both effective and rapidly deployable. There are a number of programs currently in development that are aimed at addressing the TBM threat, such as the Army Theater High-Altitude Area Defense (THAAD) system. The THAAD system will have an "upper-tier" capability to intercept missiles farther away from their targets, in the mid-course phase. However, like Patriot, THAAD will have a large deployment footprint. Although the program has encountered problems, its recent successful intercept test indicates it could become operational sometime after 2004. Another missile defense system in development is the Navy Theater Wide System, also known as "Navy Upper Tier." This ship-based system could provide coverage over a wide area, similarly to THAAD. The fact that Navy Upper Tier will be ship-based presents obvious mobility advantages. Initial operating capability of the system is projected to be after 2005. Finally, the Air Force is developing the Airborne Laser (ABL), which is intended to destroy enemy missiles during flight boost phase. A live test of the concept is scheduled for 2002. If ABL were successfully developed, it would provide a mobile missile defense capability that could deploy to rear-area theater FSLs. However, the program has technical and operational challenges to overcome and is not expected to become operational for many years.

Defense against the coming generation of advanced cruise missiles is going to be difficult, especially if the missiles incorporate even elementary low-observable technology. The key challenges lie in sensors and in command and control. The Airborne Warning and Control System (AWACS) radar will need to be improved to enable it to better detect low-flying stealthy missiles. Also, ground-based Patriot missiles will need to be able to receive cues directly from the AWACS, not just from its own fire-control radar. Patriot could then cover a much greater area, up to a radius of 100–150 km, instead of just the 25-km range associated with its ground-based radar. Such an integration of ground-based air defense systems with airborne sensors will be a complicated and expensive technical challenge.[13]

Although TBMs and cruise missiles are threats to expeditionary operations, it should be kept in mind that they will not always be a limiting factor. As we have seen recently in the Balkans, there will be times when these weapons will not be in the enemy's arsenal. At other times, the risk posed by these weapons will need to be managed. This could be accomplished through a combination of prepositioning of defenses (as with deployments to SWA today), deterring the use of such weapons, and if necessary by lengthening employment planning timelines to include the time to deploy missile defenses. Nevertheless, we believe it is important to the success of the EAF concept that the Air Force look for ways to counter accurately guided TBMs and cruise missiles within the next decade.[14]

CHEMICAL AND BIOLOGICAL THREATS

Low-Intensity CB Threats

The possible contamination of an FOL's food or water is a threat that must be considered. In addition to measures such as vaccinations and the use of antibiotics, preventative measures to counter this threat rely on good intelligence, security control of the sources of food and water by the host nation, and monitoring of transportation, distribution, and food preparation. If the security of supplies is in doubt, or intelligence indicators raise concerns, it is possible to quickly transition to bottled water and prepackaged foods (MREs). To date, these measures have been sufficient. To augment safety, water supplies could be tested for the presence of common chemical agents with equipment like the M272/E1 Water Testing Kit. In addition, systems such as the Joint Chemical/Biological Agent Water Monitor (JCBAWM), a system still under development, will be able to detect a range of both chemical and biological agents.

---

[13]Gormley (1998), p. 104.

[14]Chow et al. (1998), p. 54.

Medium-Intensity CB Threats

The localized release of CB agents defines the medium intensity level for this threat. By localized, we mean that the attack falls short of contaminating the entire FOL, leaving usable living and working areas.

A number of studies have identified the limited and possibly covert usage of CB agents as a serious asymmetric threat to our forces during both deployment and employment.[15] However, there has recently been a concerted effort to increase the operability of Air Force forward bases in the face of limited CB threats. New policies and procedures have been published, including a *Chemical and Biological Defense Concept of Operations*[16] and the *Chemical-Biological Warfare Commander's Guide.*[17] These publications have established standards and procedures for base operability during and after a chemical attack, including the designation of open-air toxic free areas (TFAs) for personnel to go to rest and recuperate. New chemical detection equipment is being procured, including both hand-held detectors and stationary alarm systems. Capabilities are considerably lower against biological threats, however, although all deployable personnel are being vaccinated against the most common agent, anthrax.[18] The Air Force is close to achieving a near-term capability to detect, isolate, decontaminate, and keep operating after limited chemical attacks. This capability will be achieved if the equipment currently in the pipeline reaches deployable units and planned training is instituted.

Transporting CB defense equipment (detectors, protective garments, etc.) does not seem to pose an insurmountable deployment problem, but mobility operations in a CB environment will be challenging. A number of studies have pointed out that small-scale attacks on aerial ports of debarkation (APODs) could have a serious adverse effect on

---

[15]Department of Defense (1997).

[16]Department of the Air Force (1998a).

[17]Department of the Air Force (1998b).

[18]It is hard to posit a limited biological attack. Most attacks of this type would be basewide (see high-intensity subsection below).

deployment operations.[19] There are currently no standards or procedures for the decontamination of Air Force airlift aircraft. Chow et al. recommend that an airlift concept of operations be developed for the CB environment.[20] It would likely require the transshipment of cargo between "clean" intertheater aircraft and dedicated "dirty" intratheater aircraft at an intermediate transfer base, resulting in a substantial slowdown in deployment.

High-Intensity CB Threats

We define the high-intensity CB threat level as an attack that causes basewide contamination by CB agents. At the medium-intensity level, nearby open-air TFAs were considered a feasible response to limited attacks. With basewide contamination, however, open-air TFAs would have to be located too far away from the FOL to maintain base operability. In this environment, collective protection (COLPROs) shelters located in or near the contaminated area would be necessary to keep the FOL in operation. COLPROs have air locks and positive pressure to secure a clean area in which personnel can rest without being in protective garments. Figure 4.1 shows a M28 deployable COLPRO inflated inside a room in a building. Limited quantities of a modified M28 shelter are being procured by the Air Force that can be set up inside the TEMPER[21] tents in use for deployments of AEF force packages.

We estimate that at least ten C-17 sorties would be required to transport deployable M28 COLPROs for a base population of 3000.[22] Although it is not inconceivable that M28s could be deployed with an AEF force package, it is more likely that these shelters would be prepositioned at an FSL or FOL with the TEMPER tents and other base support equipment. Prepositioning of collective protection is particularly desirable at Category 1 FOLs where significant infrastructure investments have been made. Operation could continue and

---

[19]Department of Defense (1997), p. 13.

[20]Chow et al. (1998).

[21]Tent, extendible, modular, personnel.

[22]This would be an additional increment of support requirements over and above the baseline AEF deployment we used in our logistics analysis.

RAND*MR1113-4.1*

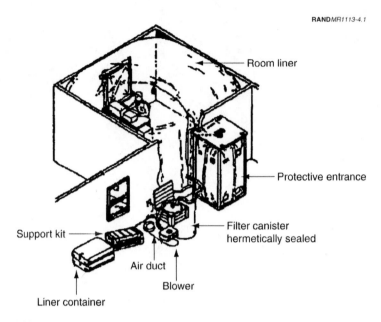

SOURCE: Department of the Army, *NBC Protection,* Field Manual 3–4, Washington, D.C., May 1992, p. 6–2.

Figure 4.1—M28 Deployable COLPRO Shelter

a time-consuming and chaotic redeployment in the midst of a contingency would be prevented. Expeditionary aerospace forces deployed forward against hostile regional powers either need access to collective protection against chemical and biological weaponry or need to be kept out of the range of the enemy's CB weapon delivery systems.

INFORMATION WARFARE THREATS

Low-Intensity Information Threats

Physical attacks on information systems represent the low-intensity end of the information warfare spectrum. If the responses to ground, air, and CB attacks are effective, Air Force information systems at FOLs should be physically secure. Because attacks that target key nodes like the wing operations center (WOC) or combat communications squadron would probably be launched locally and be ground-based, the same systems and capabilities that would protect FOLs from ground threats, such as sensors and mobile security forces, would provide for the physical security of base information systems.

Medium-Intensity Information Threats

In this part of the matrix we placed (1) threats that deny expeditionary forces the use of reachback to data and planning capabilities at other sites, (2) disruption of links with command, control, communications, computers, intelligence, surveillance, and reconnaissance (C4ISR) forces, and (3) "hacking" into Air Force systems. By cutting off expeditionary forces from their sources of C2 and targeting information, an enemy could lessen force effectiveness.

Protecting information flows is especially important for smaller applications of force, such as AEF force package deployments. In small operations, timing is critical, political sensitivity high, and the situation often dynamic and opportunistic. Information resources are leveraged to enhance the effectiveness of such forces. The jamming of intratheater communications and navigational signals is therefore a concern for deployed expeditionary forces,[23] could deny commanders the situational awareness they need to respond to events, and degrade the effectiveness and precision of force application.

High-Intensity Information Threats

Large discharges of radio waves can disrupt or damage computers and other sensitive electronics.[24] Weapons based on this principle were originally developed by the Soviet Union, and have apparently been purchased by a number of Third World countries. They could be delivered by bombs or missiles, or as mortar shells from outside the base perimeter. At close range, the weapons can be lethal; over areas hundreds of meters in diameter they can disable electronic components in computers, vehicles, and communications gear, as

---

[23]Buchan (1998).

[24]Maze (1998), p. 28.

well as aircraft. The potential of such weapons to disrupt EAF operations as well as their availability to adversaries need further study.

EAF FORCE PROTECTION ASSESSMENT

Although a broad range of threats to deployed expeditionary aerospace forces is being addressed, improvements in force protection must continue to achieve the high level of deployment flexibility needed by the Air Force. We found that the following force protection initiatives will be important to the future success of the EAF concept:

- Better sensors are needed to detect and evaluate potential ground, air, and CB threats.

- Security forces need advanced counterbattery and countersniper weapons to better respond to threats from outside base perimeters.

- An effective defense against accurate theater ballistic missiles and cruise missiles is needed within the next 10 years. Without such defensive capabilities, expeditionary aerospace forces will have to conduct operations from greater distances, outside the range of these threats.

- If AEF force packages are to have a robust capability to operate in the face of future attacks by CB weapons, they will need deployable collective protection shelters (COLPROs).

- Finally, the Air Force needs to actively investigate and evaluate potential threats to its use of reachback.

Historically, the Air Force has given little sustained attention to force protection issues. Although this changed in 1996, efforts must continue to develop force protection capabilities that cover the full spectrum of credible threats. A lack of such capabilities would substantially constrain Air Force access to bases and undermine the flexbasing strategy. For this reason, force protection will be as much of a key enabling capability as C4ISR, SEAD, and aerial refueling have been for the Air Force in the past. Without a capability to manage the risk associated with a range of threats, the Air Force will be driven to greater reliance on long-range concepts of operation using bombers and cruise missiles. As we discussed in Chapter Two, such capabilities are important, but a workable base access strategy should strike a balance between short-range and standoff employment options. A robust force protection capability will work together with agile logistics and operational flexibility to provide the EAF concept with global presence.

Chapter Five

CONCLUSIONS

Today the United States has entered an era in which both its strategic options and its challenges have broadened. The number of regions in which U.S. forces are likely to operate has multiplied, and the nature, scope, and range of political objectives are more varied as well. Although the world has always held diverse dangers, the United States today has greater latitude to act upon threats to its interests and values. At the same time, access to real-time, detailed information on global events by publics has created heightened expectations, often resulting in demands that "something be done," and be done quickly.

The effect of these developments on the U.S. Air Force has been profound. We have observed that there has been an increase in the number and duration of deployments by Air Force personnel to austere forward locations, supported by a downsized and malpositioned overseas basing structure. As a Cold War force, the Air Force was structured to fight a known enemy from well-equipped main operating bases. In the current environment, the Air Force has found it difficult to maintain high levels of readiness in the face of repeated deployments away from its home bases as well as demands for sustained presence at austere forward locations. The new operating environment requires a more agile force that can deploy rapidly and regularly, while maintaining the quality and training of its personnel.

In response to these shifting requirements, the Air Force has been restructuring itself into an EAF. Although the service has made progress in instituting the organizational aspects of moving to an EAF, we believe that there are additional concerns that need to be addressed if the Air Force is to become truly expeditionary. The Air Force needs a strategy for deploying and employing its forces overseas in the face of significant uncertainty regarding its operating locations. We have proposed such a strategy, which we call flexbasing, as a way to manage the "base access issue." However, access is not the only reason the Air Force needs a basing strategy. Such a strategy is also needed to provide expeditionary aerospace forces with the operational capability they need to perform their mission. To conclude our discussion of the flexbasing strategy, we will bring together the issues we have discussed in the previous chapters to show how the flexbasing strategy would affect the employment of expeditionary aerospace forces.

FLEXBASING, AGILE LOGISTICS, AND FORCE PROTECTION

In this report, we deemed two flexbasing elements to be especially vital to the success of the strategy: agile logistics and a robust force protection capability. Figure 5.1 conceptually illustrates the close relationship that exists between these two capabilities and the basing

RAND*MR1113-5.1*

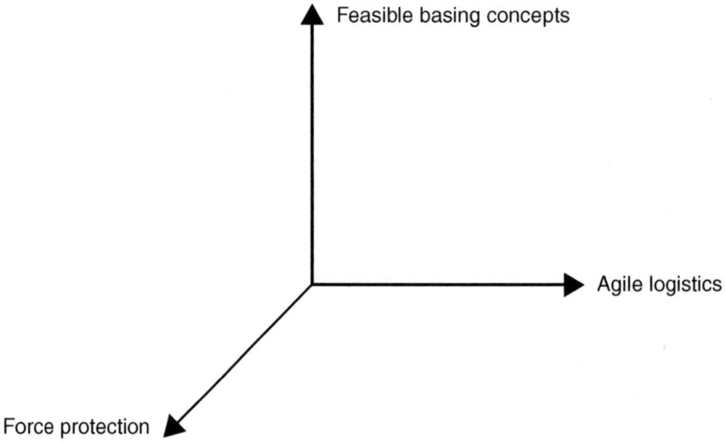

Figure 5.1—Agile Logistics, Force Protection, and Basing

concepts available to an expeditionary force. The levels of agile logistics support and the capability to operate from bases in the face of threats largely determine the range of bases that can support expeditionary operations. The basing in turn indicates whether long- or short-range weapon systems must be used, a factor in determining the intensity of the combat operations that can be brought to bear. By making available a wide range of basing alternatives, agile logistics and force protection will provide expeditionary aerospace forces the flexibility that they need to take advantage of basing opportunities. This will significantly enhance both access and the array of operational concepts open to joint commanders.

BOMBS ON TARGET: FLEXBASING AND OPERATIONAL EFFECTIVENESS

As Figure 5.1 illustrates, agile logistics and force protection are parameters that will enable the operational capability of expeditionary aerospace forces. We used these two elements of flexbasing, along with another salient aspect—FSLs—to gauge the degree to which the strategy might enhance the operational effectiveness of expeditionary aerospace forces. In Chapter Two, we described FSLs as locations where parts and equipment could be located to support operations throughout a region. They could also provide basing for longer-range combat and enabling systems such as bombers, tankers, AWACS, and the Joint Surveillance and Tracking System (JSTARS). In examining the effect of FSLs, we considered how their presence in a theater would affect the number of targets that could be struck over time. To do this, we looked at strike operations conducted with and without access to FSLs, while varying our assumptions about the levels of available logistics support and needed force protection capability.

We varied the force protection challenge according to the low, medium, and high levels that we used in our earlier force protection discussion. An increasing threat level requires substantial increases in the force protection resources that must be prepositioned at the forward location or transported to the location before operations can be safely commenced. We also varied the levels of logistics support capability using the three categories used earlier. Category 1 bases can commence operations within 48 hours of the receipt of a deployment order by CONUS-based forces. Category 2 and 3 bases provide 96-hour and 144-hour responsiveness, respectively.[1] Similar to the force protection levels, and as discussed in Chapter Three, there were significant differences in the resources that must be prepositioned or transported to meet the indicated responsiveness. Increasing levels of responsiveness require more resources, located at more bases throughout the theater of operations. Figure 5.2 shows the three-by-three matrix that results when the levels of force protection and logistics support are matched.

We examined the effects of logistics support, force protection, and the flexbasing strategy on the generation of operational capability at forward locations by looking at the three cases shown in the matrix. These cases span the ranges of threats and responsiveness, and are useful examples of the degree to which flexbasing could affect

RAND*MR1113-5.2*

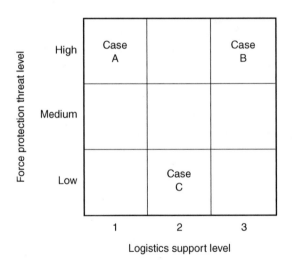

Figure 5.2—Force Protection/Logistics Matrix

---

[1]These baseline responsiveness levels assume a low force protection threat. Timelines to Initial Operational Capability (IOC) of a forward base will be extended beyond these baseline values, depending on the nature of the threat and the required responses.

operations under varying assumptions. The first, Case A, is a well-stocked, highly responsive Category 1 forward location with a high force protection threat. Case B is a Category 3 bare base, also under heavy threat. Case C is a location with 96-hour responsiveness and a low force protection threat level. We assumed that enemy TBMs with CB warheads represented the high threat level, and that the Category 1 FOL already had COLPROs prepositioned; the Category 3 FOL required both COLPROs and a Patriot battery to be airlifted to the site. For each, force protection measures had to be in place or deployed to the FOL prior to the arrival of the force package. Table 5.1 summarizes the characteristics of each case.

The mission is to conduct punitive or coercive strikes requiring the delivery of ordnance on ground targets in a remote region as quickly as possible. For the deploying forces, we used the baseline fighter aircraft package we described in Chapter Three,[2] with the addition of six B-1 and six B-2 bombers from CONUS. With these assumptions, we looked at the number of air-to-ground Joint Direct Attack Munition (JDAM) strikes that could be accomplished over a two-week period with and without access to an FSL. Figure 5.3 shows the employment concepts used with the two basing alternatives.

Table 5.1

Case Characteristics

Characteristic	Case A	Case B	Case C
Support level	Cat 1 (48 hours)	Cat 3 (144 hours)	Cat 2 (96 hours)
Threat level	High (TBMs w/CB)	High (TBMs w/CB)	Low (relatively secure)
Additional force protection requirements	Patriot battery (60 hours)	Patriot battery and COLPROs (108 hours)	No additional requirements
Time to IOC	108 hours	252 hours	96 hours

---

[2]This package consisted of 12 F-15Cs, 12 F-15Es, and 12 F-16CJs, flying 2.3, 2.3, and 2.0 sorties per day, respectively.

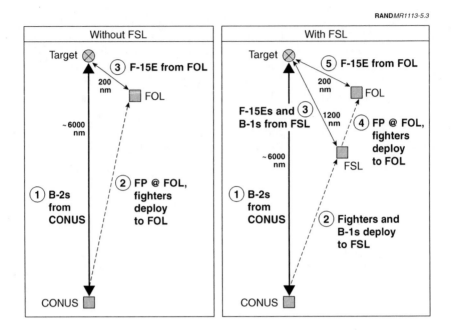

Figure 5.3—Employment Concepts With and Without an FSL

Referring to the figure, without an FSL in the theater, operations are initiated with long-range B-2 strikes from CONUS (1). At the same time, the force protection (FP) measures that are needed for Cases A and B are moved into the FOL, along with any additional required base operating support. When the base support infrastructure and force protection are ready, the fighter aircraft package deploys to the FOL (2) and commences operations (3).

With an FSL, operations similarly begin with B-2 strikes from CONUS (1). However, the fighters are deployed forward immediately to the FSL, along with the B-1s (2). While the FOL is being prepared, the B-1s begin operations from the FSL, along with the fighters that are force extended with tankers from the FSL (3). When the FOL is ready to receive the fighter package, the fighters move farther forward (4) and begin intensive operations from that location (5).

The results of our operational analysis are shown in Figure 5.4. As one would expect, with an additional base (such as an FSL) more bombs can be dropped in each case. However, the FSL appears to enable launch of intense strikes in minimum time. In Case A, with a supposedly responsive Category 1 base under high force protection threat, only 168 JDAMs can be delivered within the first five days, as opposed to 585 with an FSL. The effect of the FSL is even more evident in Case B, in which 96 JDAMs can be delivered without a theater FSL versus 561 with an FSL. This represents the difference between a

RANDMR1113-5.4

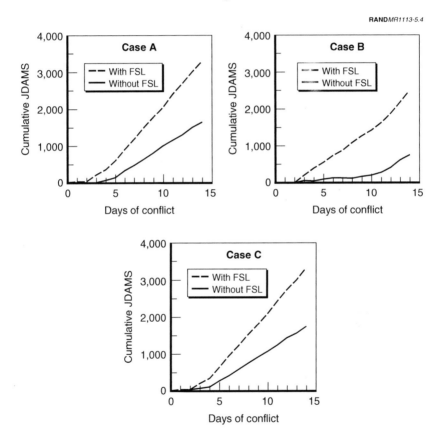

Figure 5.4—Precision-Guided Munitions Delivered

token series of strikes and the shock and intensity of a true air campaign. Case B involves movement onto a bare base, so it is evident that FSLs are especially important to expeditionary operations in theaters without an infrastructure of bases. Figure 5.5 summarizes the results over the entire 14-day period. We found that the contribution that FSLs could make to the quick response and intensity of expeditionary air campaigns is substantial.

A STRATEGY FOR GLOBAL AEROSPACE PRESENCE

We believe that flexbasing could both address base access issues and enhance the operational effectiveness of deployed expeditionary aerospace forces. In Chapter Two, we recommended that the Air Force take the following actions to implement the flexbasing strategy:

- Establish a global support infrastructure of CSLs, FSLs, and FOLs.

- Develop and maintain a robust mix of long- and short-range combat systems.

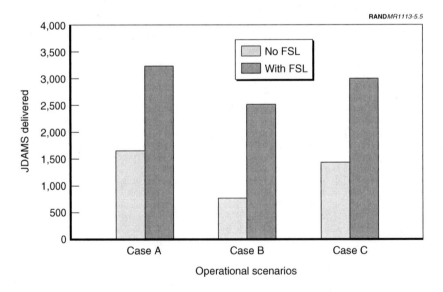

Figure 5.5—JDAMs Dropped Over 14-Day Period

- Develop earth orbit as an FSL for selected enabling capabilities.

- Advocate a global presence strategy such as flexbasing as an initiative in the joint arena.

- Design and establish a global logistics/mobility support system for expeditionary operations.

- Provide deployable full-spectrum force protection options for expeditionary forces.

Regarding logistics and mobility, we found that rapid and sustained response by expeditionary aerospace forces cannot be achieved without significant levels of regional prepositioning at FSLs or FOLs. In addition, the expeditionary logistics/mobility system that is an important part of flexbasing must be globally planned and centrally managed both to achieve the needed levels of support at minimum cost and to optimize the transportation network that will tie the system together.

To provide the robust force protection that expeditionary aerospace forces will need, we found that deployed security forces need better sensors and firepower, especially counterbattery and countersniper capabilities. We also determined that the Air Force needs to consider how it intends to address the TBM and CB threats. The responses to these threats will undoubtedly be theaterwide and grounded in the joint arena.

The locations that become available to expeditionary aerospace forces could be distant from the fight or quite close. They could be allied military bases, international airports, or abandoned airfields. To be expeditionary, the Air Force must be prepared to employ effectively from all of these locations and more, many of which will be less than ideal. Aspects such as the level of in-place support will affect the speed with which operations can begin. Locations that are distant from targets will require the employment of long-range weapon systems and greater use of aerial refueling. Locations that are close will present force protection challenges. The initiatives we have described will provide the Air Force a flexible capability to deploy to locations with a wide range of characteristics. This will reduce dependence on access to particular bases and provide potential access to many, perhaps hundreds of, locations throughout a theater of operations. Rather than focus on gaining an *a priori* assured access to specific bases, the flexbasing strategy provides a robust and flexible *capability* to move swiftly into, and operate effectively out of, whatever locations become available during crises.

REFERENCES

Buchan, Glen, et al., *Potential Vulnerabilities of Air Force Information Systems: Final Report* (U), RAND, MR-816-AF, forthcoming SECRET (no classified data used in this report.)

Buckingham, Colonel Larry A., "820th Security Forces Group," briefing, Lackland Air Force Base, Texas: 820th SFG. 1998.

Chow, Brian G., G. S. Jones, I. Lachow, J. Stillion, D. Wilkening, and H. Yee, *Air Force Operations in a Chemical and Biological Environment*, RAND, DB-189/1-AF, 1998.

Cohen, William S., Secretary of Defense, *Report of the Quadrennial Defense Review*, Washington, D.C.: Department of Defense, 1997.

Conner, Darryl, "Differences Between AEF IV, AEF V, AEF Phoenix Scorpion," point paper, Scott Air Force Base, Illinois: HQ AMC TACC/XOP, 1997.

Correll, John T., "Strung Out," *Air Force Magazine*, September 1998.

Davis, Richard, *Overseas Military Presence: More Data and Analysis Needed to Determine Whether Cost-Effective Alternatives Exist*, Washington, D.C.: General Accounting Office, GAO/NSIAD-97-133, 1997.

Department of Defense, *Assessment of the Impact of Chemical and Biological Weapons on Joint Operations in 2010 (The CB 2010 Study)*, Washington, D.C.: Booz, Allen, and Hamilton, 1997.

Department of Defense, *Joint Doctrine for Rear Area Operations*, Washington, D.C.: Joint Staff, JP 3-10, 1996.

Department of the Air Force, "50 & 5," Langley Air Force Base, Virginia: HQ ACC/HO, 1997.

Department of the Air Force, *Air Base Defense Collective Skills*, Washington, D.C.: HQ AFSPA/SPSD, AFH 31-302, 1996.

Department of the Air Force, *Chemical and Biological Defense Concept of Operations*, Tyndall Air Force Base, Florida: HQ AFCESA/CEX, 1998a.

Department of the Air Force, *Chemical-Biological Warfare Commander's Guide*, Tyndall Air Force Base, Florida: HQ AFCESA/CEX, AFP 32-4019, 1998b.

Department of the Air Force, *United States Air Force Statistical Digest*, Washington, D.C.: Assistant Secretary of the Air Force (Financial Management and Comptroller), 1992.

DFI International, *Today's Air Force*, Washington, D.C., 1997.

Fuchs, Ronald P., *Report on United States Air Force Expeditionary Forces*, Pentagon, Washington, D.C.: United States Air Force Scientific Advisory Board, SAB-TR-97-01, 1997.

Futrell, Robert F., *Ideas, Concepts, Doctrine: Basic Thinking in the United States Air Force: 1907–1960*, Maxwell Air Force Base, Alabama: Air University Press, 1989.

Gormley, Dennis M., "Hedging Against the Cruise-Missile Threat," *Survival*, Spring 1998.

Graham, Bradley, "Military Readiness, Morale Show Strain," *Washington Post*, August 13, 1998.

Harkavy, Robert E., *Bases Abroad: The Global Foreign Military Presence*, Oxford, UK: Oxford University Press, 1989.

Harkavy, Robert E., *Great Power Competition for Overseas Bases: The Geopolitics of Access Diplomacy*, New York: Pergamon Press, 1983.

Headquarters, USAF, *Expeditionary Aerospace Force Factsheet*, June 1999.

Hunter, Major Charles R., "Air Force Deployment Process and Strategy," briefing, the Pentagon, Washington, D.C.:   HQ USAF/ILXX, 1998b.

Hunter, Major Charles R., "Deliberate Crisis Action Planning and Execution System (DCAPES)," briefing, the Pentagon, Washington, D.C.: HQ USAF/ILXX, 1998a.

Kennedy, Paul M., *The Rise and Fall of British Naval Mastery*, London: Ashfield Press, 1983.

Killingsworth, Paul S., *Force Protection for Air Expeditionary Forces: Threats, Responses, and Issues (U)*, RAND, DB-263-AF, 1999. CONFIDENTIAL (no classified data used in this report.)

Kugler, Richard L., *Changes Ahead: Future Directions for the U.S. Overseas Military Presence*, RAND, MR-956-AF, 1998.

Luttwak, Edward N., *The Grand Strategy of the Roman Empire from the First Century A.D. to the Third*, Baltimore: Johns Hopkins University Press, 1976.

Maze, Richard, "Protection Sought Against Radio-Wave Interference," *Air Force Times*, March 16, 1998.

Ryan, Michael E., General, USAF, "Expeditionary Aerospace Force: A Better Use of Aerospace Power for the 21st Century," briefing, Washington, D.C.: HQ USAF, 1998.

Shlapak, David A., and Alan Vick, *Check Six Begins on the Ground, Responding to the Evolving Ground Threat to U.S. Air Force Bases*, RAND, MR-606-AF, 1995.

Stillion, John, and David Orletsky, *Airbase Vulnerability to Conventional Cruise and Ballistic Missile Attack: Technology, Scenarios, and USAF Responses*, RAND, MR-1028-AF, 1999.

Tilford, Earl H., Jr., "The 'New Look' and the Air Force," *Strategy, Doctrine, and Air Power, Book 2, Lessons 9–12*, 8th Edition, Maxwell Air Force Base, Alabama: Air War College, 1997.

Tripp, Robert S., Lionel Galway, Paul Killingsworth, Eric Peltz, Timothy Ramey, and John G. Drew, *Supporting Expeditionary*

Aerospace Forces: An Integrated Strategic Agile Combat Support Planning Framework, RAND, MR-1056-AF, 1999.

Welch, General Larry D., *Report of the Panel to Review Long-Range Air Power* (U), Washington, D.C.: Department of Defense, 1998. SECRET (no classified data used in this report.)

Wohlstetter, A. J., et al., *Selection and Use of Strategic Air Bases*, RAND, R-266, 1954.